The Observer's Pocket Series

AIRCRAFT

About the Book

The *Observer's Book of Aircraft* is the indispensable annual pocket guide to the world's latest aeroplanes and most recent versions of established aircraft types. This, the thirty-first annual edition, embraces the latest fixed- and variable-geometry aeroplanes and rotorcraft of sixteen countries. Its scope ranges from such major 1982 débutantes as the Boeing 757 short-to-medium range airliner and Saab-Fairchild SF 340 regional airliner, through important new derivative types, such as the Airbus A310 airliner, F-5G Tigershark and F-16E fighters, the SF.260-TP, Tangará and Pillan trainers, and the Metro IIIA and Commuter 1900 regional transports, to the latest versions of a wide variety of well-established aircraft types. All data has been checked and, where necessary, updated for this volume, and a large proportion of the three-view silhouettes depicting types that have appeared in previous editions have been revised to reflect the latest changes introduced by their manufacturers, or, in the case of Soviet aircraft, the most recently available information.

About the Author

William Green, compiler of the *Observer's Book of Aircraft* for 31 years, is internationally known for many works of aviation reference. William Green entered aviation journalism during the early years of World War II, subsequently served with the RAF and resumed aviation writing in 1947. He is currently managing editor of one of the largest-circulation European-based aviation journals, *Air International*, and co-editor of *Air Enthusiast* and the *RAF Yearbook*.

The Observer's Book of

AIRCRAFT

COMPILED BY
WILLIAM GREEN

WITH SILHOUETTES BY
DENNIS PUNNETT

DESCRIBING 142 AIRCRAFT
WITH 247 ILLUSTRATIONS

1982 EDITION

FREDERICK WARNE

© FREDERICK WARNE & CO LTD
LONDON, ENGLAND
1982

Thirty-first edition 1982

LIBRARY OF CONGRESS CATALOG CARD NO: 57 4425

ISBN 0 7232 1618 5

Printed in Great Britain

_1676.1281_

INTRODUCTION TO THE 1982 EDITION

This year of 1982 has dawned to find most of the world's major airlines and, in consequence, the suppliers of their equipment, in critical situations. Many prominent operators have been metaphorically brought to their knees, with all the big aircraft manufacturers suffering as a result, and there is little sign of any easing of the recession in aircraft sales. Before last year's end, one of the largest components of the US industry, Lockheed, had announced that it was cutting its losses and terminating TriStar production from 1984; the production future of the competitive McDonnell Douglas DC-10 hangs by the slenderest of threads and demand for new Boeing 747s has diminished alarmingly. The lack of buoyancy in the large airliner market is being felt, too, in the corporate executive and regional transport markets, with most manufacturers paring their production plans for this year.

On the military scene, the Rockwell B-1 strategic bomber has been resurrected and will resume flight test next year. Conversely, there seems at the time of closing for press every likelihood that production of the perennial P-3 Orion maritime patrol aircraft for the US Navy will terminate in 1983, rather than in 1989 as previously envisaged, and budgetary strictures on the other side of the Atlantic are necessitating cutbacks in this year's planned production rate of the trinational Tornado combat aircraft.

These gloomy results of international recession are reflected in the *status* entries for many of the aircraft types appearing in this, the thirty-first annual edition of *The Observer's Book of Aircraft*. Another aspect of this year's edition is the paucity of aircraft débutantes in its pages, the most important of the very few new types expected to make their first flights during the currency of this volume being the Boeing 757 short-to-medium range airliner, which will hopefully be joined in the air before year's end by the Saab-Fairchild SF 340, the first of a new generation of turboprop-powered regional transports.

If genuine débutantes are few and far between in this edition, however, not so the new derivative aircraft appearing for the first time. These include the Airbus A310, the F-5G Tigershark, the Lockheed TR-1, the Siai Marchetti SF.260-TP and SF.600-TP, the Fairchild Swearingen Metro IIIA, the Aerotec Tangará, Beechcraft's Commuter C99 and 1900, the Piper Pillan, the IAI Kfir-TC2 and the General Dynamics F-16E. The BAe 146, Boeing 767, Shorts 360, AV-8B Harrier II and Peregine have all entered flight test since the last edition appeared, and it is finally possible to illustrate the immense Mi-26 helicopter, which has, in fact, been flying since 1977.

WILLIAM GREEN

AERITALIA AP 68B

Country of Origin: Italy.

Type: Light multi-role transport and utility aircraft.

Power Plant: Two (flat-rated) 330 shp Allison 250-B17C turboprops.

Performance: (AP 68B-100) Max. cruising speed, 240 mph (385 km/h) at 15,000 ft (4 570 m); econ cruise, 190 mph (306 km/h) at 18,000 ft (5 475 m); initial climb, 2,025 ft/min (10,29 m/sec); service ceiling, 25,000 ft (7 620 m); range (max. payload), 358 mls (576 km) at 235 mph (378 km/h) at 10,000 ft (3 050 m); max. range, 1,045 mls (1 682 km) at 181 mph (291 km/h).

Weights: Empty, 3,245 lb (1 472 kg); max. take-off, 5,732 lb (2 600 kg).

Accommodation: Basic arrangement for eight persons (including pilot) in four side-by-side pairs.

Status: The prototype (AP 68TP) was flown on 11 September 1978, and the first of an initial series of 25 aircraft (AP 68B-100) was flown 20 November 1981, with customer deliveries scheduled for mid 1982.

Notes: Based on the Partenavia P 68C, the AP 68B is intended for both civil and military roles, the Series 100 (described and illustrated opposite) having a fixed undercarriage and the Series 200 having a retractable undercarriage similar to that of the prototype (illustrated above).

AERITALIA AP 68B-100

Dimensions: Span, 39 ft 4½ in (12,00 m); length, 31 ft 10 in (9,70 m); height, 11 ft 11¾ in (3,65 m); wing area, 200·21 sq ft (18,60 m²).

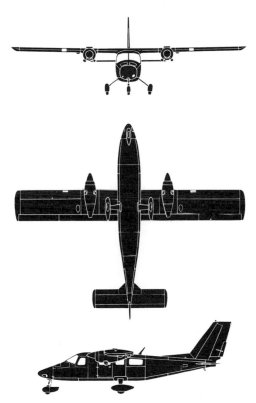

AERITALIA G.222T

Country of Origin: Italy.

Type: General-purpose military transport.

Power Plant: Two 4,860 shp Rolls-Royce Tyne RTy 20 (Mk. 801) turboprops.

Performance: Max. speed (at 54,012 lb/24 500 kg), 334 mph (537 km/h) at 29,000 ft (8 840 m), 350 mph (563 km/h) at 20,000 ft (6 095 m); long-range cruise, 265 mph (426 km/h) at 20,000 ft (6 095 m); range (with 10,000 lb/4 536 kg payload), 1,324 mls (2 130 km) at long-range cruise; ferry range (no payload), 2,880 mls (4 635 km); service ceiling, 25,000 ft (7 620 m).

Weights: Operational empty, 40,500 lb (18 371 kg); max. take-off, 61,730 lb (28 000 kg).

Accommodation: Flight crew of three–four and 53 fully-equipped troops, 42 paratroops or 36 casualty stretchers plus four medical attendants. Palletised or non-palletised freight loads up to 18,740 lb (8 500 kg).

Status: Prototype G.222T (conversion of 34th production G.222 airframe) flown 15 May 1980, with initial order for 20 from Libya with first delivered on 20 February 1981, and six–seven delivered by beginning of 1982.

Notes: The G.222T is a Tyne-engined version of the basic G.222 (see 1978 edition), which is powered by General Electric T64-P4D turboprops, and production of which continues in parallel. Electronic warfare, navigational aid calibration and maritime surveillance versions are under development.

AERITALIA G.222T

Dimensions: Span, 94 ft 2 in (28,70 m); length, 74 ft 5½ in (22,70 m); height, 32 ft 1¾ in (9,80 m); wing area, 882·64 sq ft (82,00 m²).

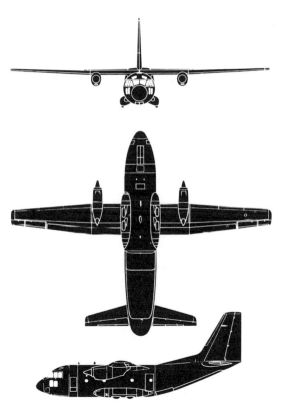

AERMACCHI MB-339A

Country of Origin: Italy.

Type: Tandem two-seat basic/advanced trainer.

Power Plant: One 4,000 lb st (1 814 kgp) Fiat-built Rolls-Royce Viper 632-43 turbojet.

Performance: Max. speed, 558 mph (898 km/h) at sea level, 508 mph (817 km/h) at 30,000 ft (9 145 m); initial climb, 6,595 ft/min (33,5 m/sec); time to 30,000 ft (9 145 m), 7·1 min; service ceiling, 48,000 ft (14 630 m); max. range (internal fuel with 10% reserves), 1,093 mls (1 760 km), (with two drop tanks), 1,310 mls (2 110 km).

Weights: Empty, 6,780 lb (3 075 kg); normal loaded, 9,700 lb (4 400 kg); max. take-off, 13,000 lb (5 897 kg).

Armament: For armament training and light strike roles a max. of 4,000 lb (1 815 kg) may be distributed between six underwing stations when flown as a single-seater.

Status: First of two prototypes flown 12 August 1976, with first of six pre-series aircraft following on 20 July 1978. First deliveries to Italian Air Force February 1981 against requirement for 100 aircraft, preceded by first export delivery (10 for Argentine Navy) commencing November 1980, followed by deliveries to the Peruvian Air Force (14 aircraft) in second half of 1981.

Notes: MB-339A is currently being manufactured in parallel with a single-seat light close air support version, the MB-339K Veltro 2 (see 1981 edition) first flown 30 May 1980. Production of the MB-339K was initiated in March 1981. The licence assembly of the MB-339A is expected to be undertaken in Peru, the Peruvian Air Force having a requirement for 66 aircraft.

AERMACCHI MB-339A

Dimensions: Span, 35 ft 7 in (10,86 m); length, 36 ft 0 in (10,97 m); height, 13 ft 1 in (3,99 m); wing area, 207·74 sq ft (19,30 m²).

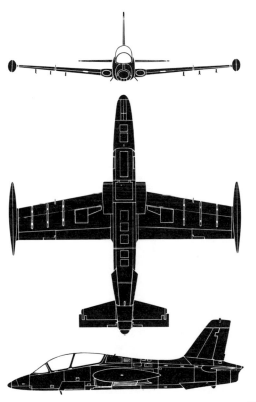

AÉROSPATIALE TB 30 EPSILON

Country of Origin: France.

Type: Tandem two-seat primary-basic trainer.

Power Plant: One 300 shp Avco Lycoming IO-540-L1B5-D six-cylinder horizontally-opposed engine.

Performance: (At 2,645 lb/1 200 kg) Max. speed, 238 mph (383 km/h) at 6,000 ft (1 830 m); max. cruise, 230 mph (370 km/h) at sea level; initial climb, 1,700 ft/min (8,63 m/sec); service ceiling, 20,010 ft (6 100 m); endurance (at 50% power), 3·75 hr at 5,020 ft (1 530 m).

Weights: Empty equipped, 1,936 lb (878 kg); max. take-off, 2,645 lb (1 200 kg).

Status: First prototype Epsilon flown on 22 December 1979, with second following on 12 July 1980. Budgetary provision made in May 1981 for initial batch of 30 for Armée de l'Air, with second batch of 30 provided for in Fiscal 1982 funding. First deliveries scheduled for late 1982 or early 1983.

Notes: Scheduled to enter service at the Armée de l'Air college at Salon de Provence mid-'83, the Epsilon is expected to be utilised for 80–100 hours flying training after grading on the CAP 10. The cockpit design, manœuvre performance and landing speed are claimed to have been matched as closely as possible to those of the Alpha Jet, to which Armée de l'Air pupils will progress, and the possibility of installing lightweight ejection seats in the production model was under investigation at the beginning of 1982. Since the late 1979 commencement of flight testing, the Epsilon has undergone numerous modifications, including redesign of the rear fuselage and vertical tail surfaces and the application of extended and upturned wingtips, the anticipated production configuration being shown by the drawing opposite.

12

AÉROSPATIALE TB 30 EPSILON

Dimensions: Span, 25 ft 11½ in (7,92 m); length, 24 ft 10½ in (7,59 m); height, 8 ft 8¾ in (2,66 m); wing area, 103·34 sq ft (9,60 m²).

AEROTEC A-132 TANGARÁ

Country of Origin: Brazil.

Type: Side-by-side two-seat primary trainer.

Power Plant: One 160 hp Avco Lycoming O-320-B2B four-cylinder horizontally-opposed engine.

Performance: Max speed, 148 mph (238 km/h) at sea level; cruise (75% power), 121 mph (195 km/h), (60% power), 108 mph (174 km/h); initial climb, 905 ft/min (4,6 m/sec); service ceiling, 14,760 ft (4 500 m); endurance (with 30 min reserves), 4 hrs 18 min.

Weights: Empty equipped, 1,234 lb (560 kg); max. take-off, 1,896 lb (860 kg).

Status: The prototype Tangará was flown for the first time in February 1981, and orders for up to 100 for the Brazilian Air Force were anticipated at the beginning of 1982, with service entry from early 1983 as the T-17.

Notes: The Tangará is a derivative of the A-122 Uirapuru and is intended as a successor to the earlier aircraft. Unlike its predecessor, the Tangará is cleared for full aerobatics and has a revised fuselage of simplified construction, a revised and improved cockpit layout, a new aft-sliding cockpit canopy and a small increase in gross wing area. Although primarily intended for service with the Brazilian Air Force, the Tangará is also intended for use by Brazilian civil flying clubs. The service designation T-17 was adopted rather than the correct sequential designation (i.e., T-28) which would have been in advance of that of the T-27 Tucano basic trainer to which pupils will progress.

AEROTEC A-132 TANGARÁ

Dimensions: Span, 29 ft 6¼ in (9,00 m); length, 22 ft 11½ in (7,00 m); height, 8 ft 10¼ in (2,70 m); wing area, 148·22 sq ft (13,77 m²).

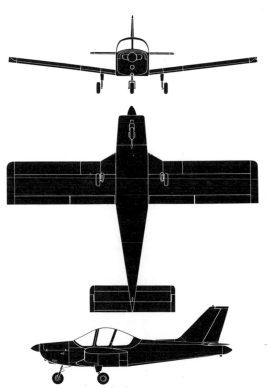

AHRENS AR 404

Country of Origin: Puerto Rico.
Type: Freighter and utility transport.
Power Plant: Four 420 shp Allison 250-B17C turboprops.
Performance: Max. continuous cruising speed, 201 mph (323 km/h) at 800 ft (244 m); range (standard fuel), 980 mls (1 577 km), (with optional auxiliary tanks), 1,382 mls (2 224 km); initial climb, 1,200 ft/min (6,09 m/sec); service ceiling, 18,000 ft (5 485 m).
Weights: Empty, 9,980 lb (4 527 kg); max. take-off, 18,500 lb (8 392 kg).
Accommodation: Flight crew of two and max. payload of 8,520 lb (3 865 kg). Up to 30 passengers may be accommodated in commuter version with 10 rows of three-abreast seating.
Status: Prototype of AR 404 (built in the USA) was flown on 1 December 1976, and production prototype (built wholly in Puerto Rico) followed on 26 October 1979. This prototype was subsequently brought up to production standards in which form it first flew on 23 September 1981. Type certification is anticipated mid-1982, and letters of intent for more than 100 aircraft were claimed at the beginning of that year when second aircraft had joined the certification test programme. Production is expected to attain one aircraft monthly by late 1982, rising to a peak of four monthly by late 1983.
Notes: Intended as a simple and robust transport suitable for operation under primitive conditions, the AR 404 has a constant-section square fuselage with manually-operated rear loading ramp, the wing being mounted on the fuselage upper surface and the mainwheels housed within sponsons.

AHRENS AR 404

Dimensions: Span, 66 ft 0 in (20,12 m); length, 52 ft 9 in (16,08 m); height, 19 ft 0 in (5,79 m); wing area, 422 sq ft (39,20 m²).

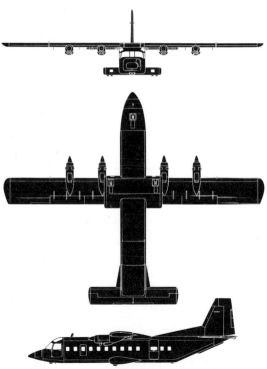

AIRBUS A300B4-200

Country of Origin: International consortium.

Type: Medium-haul commercial transport.

Power Plant: Two 52,500 lb st (23 814 kgp) General Electric CF6-50C2 (B4-203) or 53,000 lb st (24 040 kgp) Pratt & Whitney JT9D-59A1 (B4-220) turbofans.

Performance: Max. cruising speed, 552 mph (889 km/h) at 31,000 ft (9 450 m); econ. cruise, 535 mph (861 km/h) at 33,000 ft (10 060 m); long-range cruise, 530 mph (854 km/h); range (max. payload), 3,430 mls (5 520 km), (max. fuel and 61,600 lb/27 940 kg payload), 4,213 mls (6 780 km).

Weights: Operational empty, 194,900 lb (88 407 kg); max. take-off, 363,750 lb (165 000 kg).

Accommodation: Flight crew of three and various seating arrangements for 220–336 passengers in six-, seven-, eight- or nine-abreast seating.

Status: First A300B flown 28 October 1972, and first A300B4 on 26 December 1974. Production (A300B2 and B4) running at 4·25 monthly at beginning of 1982, when 255 (plus 69 options) had been ordered.

Notes: The A300B is manufactured by a consortium of Aérospatiale, British Aerospace and Deutsche Airbus, the A300B4 version described above being a longer-range version of the A300B2 and the -200 differing from the -100 in having reinforced wings and fuselage and strengthened undercarriage to cater for higher take-off weights. The A300B4-600, deliveries of which are to commence spring 1984, will feature the A310 rear fuselage coupled with an extended parallel section of the rear fuselage and the new-generation engines offered on the A310, thus offering significant commonality between the two types.

18

AIRBUS A300B4-200

Dimensions: Span, 147 ft 1¼ in (44,84 m); length, 175 ft 11 in (53,62 m); height, 54 ft 2 in (16,53 m); wing area, 2,799 sq ft (260,00 m²).

AIRBUS A310-200

Country of Origin: International consortium.

Type: Short/medium-haul commercial transport.

Power Plant: Two 48,000 lb st (21 773 kgp) General Electric CF6-80A-1 (A310-202) or Pratt & Whitney JT9D-7R4D1 (A310-221), or 50,000 lb st (22 680 kgp) CF6-80A-3 or JT9D-7R4E1 turbofans.

Performance: Max. cruising speed, 562 mph (904 km/h) at 33,000 ft (10 060 m); econ. cruise, 528 mph (850 km/h) at 37,000 ft (11 280 m); long-range cruise, 514 mph (828 km/h); range (with max. payload), 2,540 mls (4 090 km), (max. fuel and 27,100 lb/12 290 kg payload), 5,095 mls (8 200 km).

Weights: Operational empty, 169,500 lb (76 885 kg); max. take-off, 305,560 lb (138 600 kg).

Accommodation: Flight crew of two or three and maximum seating for 265 passengers nine abreast.

Status: First A310 scheduled to fly end of March 1982, with first customer deliveries spring 1983. A total of 88 (plus 90 on option) ordered by beginning of 1982.

Notes: The A310 is a short-fuselage development of the A300 and also features new, advanced technology wings of reduced span and area, a new, smaller tailplane and standardised engine mounting pylons catering for all types of power plant offered. Versions with both the General Electric and Pratt & Whitney engines are to be certificated in parallel, permitting first deliveries to both Lufthansa and Swissair to be effected simultaneously. A convertible version, the A310C, will have a large upper-deck cargo door and reinforced cabin floor.

AIRBUS A310-200

Dimensions: Span, 144 ft 0 in (43,90 m); length, 153 ft 1 in (46,66 m); height, 51 ft 10 in (15,81 m); wing area, 2,357 sq ft (219,00 m²).

ANTONOV AN-28 (CASH)

Country of Origin: USSR (Poland).

Type: Light STOL utility transport.

Power Plant: Two 960 shp PZL-Rzeszov-built Glushenkov TVD-10B turboprops.

Performance: Max. cruising speed, 217 mph (350 km/h); econ. cruise, 186 mph (300 km/h); max. climb, 2,460 ft/min (12,49 m/sec); range (at econ. cruise), 807 mls (1 300 km) with max. fuel, 410 mls (660 km) with 15 passengers, 233 mls (375 km) with 18 passengers.

Weights: Empty, 7,716 lb (3 500 kg); normal loaded, 12,785 lb (5 800 kg); max. take-off, 13,450 lb (6 100 kg).

Accommodation: Flight crew of one or two and basic arrangement (passenger version) for 15 seats three abreast, or up to 20 seats in high density configuration. An ambulance version accommodates six casualty stretchers, five seated casualties and a medical attendant. An electrically-actuated ramp beneath the rear fuselage facilitates direct loading from trucks when used in the freighter role.

Status: Initial prototype flown as An-14M in September 1969, a production prototype following early in 1974, the aircraft having meanwhile been redesignated An-28. The production programme was subsequently transferred to Poland where manufacture is now being undertaken by the PZL (the airframe at Lublin/Swidnik and the engines at Rzeszov). Deliveries are scheduled to commence in 1983–84, current planning calling for the export of 1,200 aircraft of this type to the Soviet Union by 1990, with an annual production rate of the order of 200 aircraft.

ANTONOV AN-28 (CASH)

Dimensions: Span, 72 ft 2⅛ in (22,06 m); length, 42 ft 6⅞ in (12,98 m); height, 15 ft 1 in (4,60 m); wing area, 433·58 sq ft (40,28 m²).

ANTONOV AN-32 (CLINE)

Country of Origin: USSR.

Type: Commercial freighter and military tactical transport.

Power Plant: Two 4,190 ehp Ivchenko AI-20M turboprops.

Performance: Max. continuous cruising speed, 317 mph (510 km/h) at 26,250 ft (8 000 m); service ceiling, 31,150 ft (9 500 m); range (with max. payload and 45 min reserves), 497 mls (800 km), (with max. fuel), 1,367 mls (2 200 km).

Weights: Max. take-off, 57,270 lb (26 000 kg).

Accommodation: Flight crew of five and 39 passengers on tip-up seats along fuselage sides, 30 fully-equipped paratroops or 24 casualty stretchers and one medical attendant. Up to 13,227 lb (6 000 kg) of freight may be carried, loading being facilitated by rear ramp and 4,409 lb (2 000 kg) capacity cargo hoist.

Status: The prototype An-32 was flown late 1976, and production initiated on behalf of Indian Air Force with deliveries against an order for 50 commencing late 1982 or early 1983. These are to be followed by a further 45 aircraft which will be assembled from component kits at Kanpur, India.

Notes: Based on the airframe of the An-26 (Curl) but embodying structural strengthening to absorb a 33% increase in installed power, the An-32 has been designed specifically to meet a requirement formulated by the Indian Air Force for a medium tactical transport possessing good "hot-and-high" performance characteristics with good field performance and the ability to operate from unpaved strips with minimal support. A variety of small wheeled or tracked vehicles (e.g., the Soviet GAZ-69 and UAZ-469) can be accommodated.

ANTONOV AN-32 (CLINE)

Dimensions: Span, 95 ft 9½ in (29,20 m); length, 78 ft 1 in (23,80 m); height, 28 ft 1½ in (8,58 m); wing area, 807·1 sq ft (74,98 m²).

ANTONOV AN-72 (COALER)

Country of Origin: USSR.

Type: Short-haul STOL utility transport.

Power Plant: Two 14,330 lb st (6 500 kgp) Lotarev D-36 turbofans.

Performance: Max. cruising speed, 447 mph (720 km/h); service ceiling, 36,090 ft (11 000 m); normal operating altitude, 26,250–32,800 ft (8 000–10 000 m); range (max. fuel), 2,360 mls (3 800 km), (with 16,534 lb/7 500 kg payload and 30 min reserves), 1,243 mls (2 000 km), (with 22,045 lb/10 000 kg payload), 620 mls (1 000 km).

Weights: Max. take-off (for 3,280 ft/1 000 m runway), 58,420 lb (26 500 kg), (for 4,920 ft/1 500 m runway), 72,751 lb (33 000 kg).

Accommodation: Flight crew of three and provision for up to 32 passengers on fold-down seats along cabin sides, or 24 casualty stretchers plus one medical attendant. Rear loading ventral ramp with clamshell doors for cargo hold with overhead hoist and provision for roller floor.

Status: First of two prototypes flown 31 August 1977. Completion of flight testing mid-1981 when it was announced that the An-72 would enter service during the 1981–85 Five-year Plan.

Notes: Allegedly intended primarily for use in Siberia, the Soviet Far East and Central Asia, the An-72 achieves short take-off and landing characteristics by means of upper surface blowing, engine exhaust gases flowing over the upper wing surfaces and the inboard double-slotted flaps. Capable of operation from short, semi-prepared strips, the An-72 has obvious military potentialities.

ANTONOV AN-72 (COALER)

Dimensions: Span, 84 ft 9 in (25,83 m); length, 87 ft 2½ in (26,58 m); height, 27 ft 0 in (8,24 m).

BAe 146 SERIES 200

Country of Origin: United Kingdom.
Type: Short-haul regional commercial transport.
Power Plant: Four 6,700 lb st (3 040 kgp) Avco Lycoming ALF 502R-3 turbofans.
Performance: Max. cruising speed, 482 mph (775 km/h) at 26,000 ft (7 925 m); econ. cruise, 440 mph (709 km/h) at 30,000 ft (9 145 m); long-range cruise, 436 mph (702 km/h) at 30,000 ft (9 145 m); range (max. payload), 1,768 mls (2 845 km) at econ. cruise, (max. fuel), 1,895 mls (3 050 km), or (with optional fuel capacity), 2,153 mls (3 465 km) at long-range cruise.
Weights: Operational empty, 47,190 lb (21 405 kg); max. take-off, 88,250 lb (40 030 kg).
Accommodation: Flight crew of two and maximum seating (one class) for 106 passengers in six-abreast seating.
Status: First BAe 146 (Series 100) flown 3 September 1981. Second (also Series 100) was scheduled to fly January 1982, with first Series 200 (fifth off line) to fly spring 1982. First customer delivery scheduled for late 1982, with planned production rate of three monthly by 1983. Firm orders had been placed for 13 BAe 146s (two Series 100 and 11 Series 200) by the beginning of 1982 when options had been taken on a further 12.
Notes: The BAe 146 is currently being built in two versions, the Series 100 (illustrated above) and the Series 200 (described above and illustrated on opposite page), the former having a length of 85 ft 10 in (26,16 m) and basic passenger accommodation for 82. The BAe 146 is optimised to operate over stage lengths of the order of 150 miles (240 km) with unrefuelled multi-stop capability. Apart from fuselage length and capacity, the Series 100 and 200 are similar in all respects.

BAE 146 SERIES 200

Dimensions: Span, 85 ft 5 in (26,34 m); length, 93 ft 8½ in (28,56 m); height, 28 ft 3 in (8,61 m); wing area, 832 sq ft (77,30 m²).

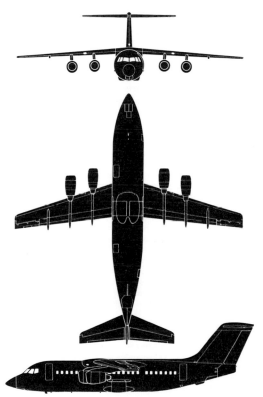

BAe HARRIER GR Mk 3

Country of Origin: United Kingdom.

Type: Single-seat V/STOL close support and reconnaissance aircraft.

Power Plant: One 21,000 lb st (9 526 kgp) Rolls-Royce Pegasus 103 vectored-thrust turbofan.

Performance: Max. speed (clean aircraft), 648 mph (1 042 km/h) or Mach 0·85 at sea level, 595 mph (957 km/h) or Mach 0·9 at 36,000 ft (10 970 m); tactical radius (LO-LO-LO mission profile with 3,000 lb/1 360 kg ordnance), 175 mls (280 km), (with two 100 Imp gal/378 l drop tanks and 1,500 lb/680 kg ordnance), 230 mls (370 km), (HI-LO-HI), 460 mls (740 km); ferry range, 2,070 mls (3 330 km); time to 40,000 ft (VTO), 2·38 min; service ceiling, 48,000 ft (14 630 m).

Weights: Operational empty, 12,140 lb (5 507 kg); max. loaded (for VTO), 18,000 lb (8 165 kg), (for STO), 26,000 lb (11 793 kg).

Armament: Provision for two 30-mm Aden cannon plus two AIM-9 Sidewinder AAMs and up to 5,000 lb (2 268 kg) ordnance on five stations.

Status: First of six pre-series aircraft flown 31 August 1966, with first of 77 GR Mk 1 version for RAF following 28 December 1967. GR Mk 1s and 13 two-seat T Mk 2s progressively brought up to GR Mk 3 and T Mk 4 standards, with follow-on orders for 37 GR Mk 3s and four T Mk 4s for the RAF and four T Mk 4s for the Royal Navy. Production of 102 Mk 50s (equivalent to GR Mk 3) and eight two-seat Mk 54s (equivalent to T Mk 4) for US Marine Corps (as AV-8As and TAV-8As), plus 11 Mk 50s and two Mk 54s for Spanish Navy (by which known as Matador), and two T Mk 60s for the Indian Navy.

Notes: Progressive development of basic design by McDonnell Douglas in collaboration with BAe has resulted in the AV-8B Harrier II (see pages 152–3).

BAe HARRIER GR MK 3

Dimensions: Span, 25 ft 3 in (7,70 m); length, 45 ft 7¾ in (13,91 m); height, 11 ft 3 in (3,43 m); wing area, 201·1 sq ft (18,68 m²).

BAe HAWK T Mk 1

Country of Origin: United Kingdom.

Type: Tandem two-seat basic/advanced trainer and light tactical aircraft.

Power Plant: One 5,200 lb st (2 360 kgp) Rolls-Royce Turboméca RT.172–06–11 Adour 151 turbofan.

Performance: Max. speed, 622 mph (1 000 km/h) or Mach 0·815 at sea level, 580 mph (933 km/h) or Mach 0·88 at 36,000 ft (10 970 m); tactical radius (HI-LO-HI mission profile with 5,600 lb/2 540 kg ordnance), 345 mls (560 km), (with 3,000 lb/1 360 kg ordnance and two drop tanks), 645 mls (1 040 km); time to 30,000 ft (9 145 m), 6·1 min.

Weights: Empty, 8,040 lb (3 647 kg); loaded (clean), 11,100 lb (5 040 kg); max. take-off, 17,085 lb (7 757 kg).

Armament: Some 90 RAF Hawk T Mk 1s were being modified at beginning of 1982 to carry a pair of AIM-9 Sidewinder AAMs to provide secondary air defence capability. Provision for four wing stores stations (for close air support) for max. ordnance load of 5,600 lb (2 540 kg).

Status: Single pre-production example flown 21 August 1974, followed by first production example on 19 May 1975. Total of 175 ordered for RAF, and export orders by beginning of 1982 were as follows: Finland (Mk 51) 50 of which 46 assembled by Valmet, Kenya (Mk 52) 12, Indonesia (Mk 53) 17, Zimbabwe (Mk 54) 8.

Notes: In November 1981, the US Navy announced selection of the Hawk as winner of its VTXTS undergraduate flight trainer. Licence manufacture of the Hawk for the US Navy will be undertaken by McDonnell Douglas against requirement for 280–324 aircraft.

BAe HAWK T Mk 1

Dimensions: Span, 30 ft 9¾ in (9,39 m); length, 38 ft 10⅔ in (11,85 m); height, 13 ft 1 in (4,00 m); wing area, 179·64 sq ft (16,69 m²).

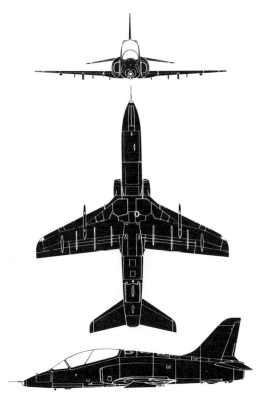

BAe HS 125 SERIES 700

Country of Origin: United Kingdom.

Type: Light business executive transport.

Power Plant: Two 3,700 lb st (1 680 kgp Garrett AiResearch TFE 731-3-1H turbofans.

Performance: Max. cruising speed, 495 mph (796 km/h) at 27,500 ft (8 380 m); econ. cruise, 449 mph (722 km/h) at 37,000–41,000 ft (11 275–12 500 m); range (max. fuel and max. payload with 45 min reserves), 2,785 mls (4 482 km); time to 35,000 ft (10 675 m), 19 min; service ceiling, 41,000 ft (12 500 m).

Weights: Empty, 12,845 lb (5 826 kg); max. take-off, 25,500 lb (11 566 kg).

Accommodation: Normal flight crew of two and basic layout for eight passengers, with alternative arrangements for up to 14 passengers.

Status: Series 700 development aircraft flown 28 June 1976, followed by first production aircraft on 8 November 1976. The 142nd Series 700 (the 500th HS 125 sold) was delivered late October 1981, and production was running at three monthly at the beginning of 1982, with the 150th Series 700 scheduled for January–February delivery.

Notes: The Series 700 differs from preceding series of HS 125s primarily in having turbofans in place of Viper turbojets, and a retrofit programme for the conversion of earlier models to TFE 731 engines is in progress with more than 50 converted, these being identified by an "F" prefix to their series number (e.g., Series F600, etc). Options for developed versions under study at the beginning of 1982 include a modest fuselage stretch and 4,000 lb st (1 814 kgp) TFE 731-5 engines.

34

BAe HS 125 SERIES 700

Dimensions: Span, 47 ft 0 in (14,32 m); length, 50 ft 8½ in (15,46 m); height, 17 ft 7 in (5,37 m); wing area, 353 sq ft (32,80 m²).

BAe HS 748 SERIES 2B

Country of Origin: United Kingdom.
Type: Regional commercial transport.
Power Plant: Two 2,280 ehp Rolls-Royce Dart RDa 7 Mk 536-2 turboprops.
Performance: Max. cruising speed, 283 mph (456 km/h) at 12,000 ft (3 660 m); long-range cruise, 269 mph (433 km/h) at 25,000 ft (7 620 m); range (with max. payload), 1,324 mls (2 130 km) at 279 mph (448 km/h) at 17,000 ft (5 180 m), (with max. fuel), 2,050 mls (3 300 km) at 269 mph (433 km/h) at 25,000 ft (7 620 m).
Weights: Operational empty, 26,000 lb (11 794 kg); max. take-off, 46,500 lb (21 092 kg).
Accommodation: Normal flight crew of two with arrangements for 44 to 58 passengers four abreast in paired seats.
Status: First prototype HS 748 flown 24 June 1960, and prototype Series 2B flown on 22 June 1979, this version superseding the Series 2A with first customer delivery in January 1980. Orders (all versions) totalled 357 at beginning of 1982, when production was rising from 1·0 to 1·5 monthly.
Notes: At the beginning of 1982, British Aerospace was investigating several projected growth versions of the HS 748 with such features as increased window pitch, advanced-technology flight deck and swept vertical tail surfaces, and various options of fuselage stretch (up to 60–64 passengers) and power plant of 2,800–3,000 hp.

BAe HS 748 SERIES 2B

Dimensions: Span, 102 ft 6 in (31,23 m); length, 67 ft 0 in (20,42 m); height, 24 ft 10 in (7,57 m); wing area, 828·87 sq ft (77,00 m²).

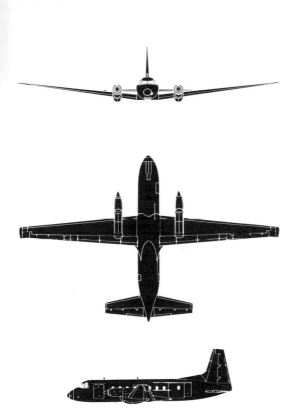

BAe JETSTREAM 31

Country of Origin: United Kingdom.
Type: Light business and utility transport.
Power Plant: Two 900 shp (flat-rated) Garrett AiResearch TPE 331-10 turboprops.
Performance: Max. cruising speed, 303 mph (488 km/h) at 16,000 ft (4 875 m); range cruise, 291 mph (469 km/h); initial climb, 2,231 ft/min (11,33 m/sec); service ceiling, 31,600 ft (9 480 m); range (six passengers with 30 min reserves plus 5%), 1,275 mls (2 053 km), (eight passengers), 1,150 mls (1 852 km).
Weights: Empty equipped (including two crew), 8,840 lb (4 010 kg); max. take-off, 14,550 lb (6 600 kg).
Accommodation: Two seats side-by-side on flight deck with basic corporate executive seating for eight passengers, with optional 12-seat (executive shuttle) arrangement and third-level commuter airliner arrangement for 18 passengers in three-abreast seating.
Status: Flight development Jetstream 31 (converted from a Series 1 airframe) flown 28 March 1980, with first production aircraft scheduled to fly February–March 1982, with first customer deliveries (in commuterliner version) in second half of year. Planned production rate of 25 per year to be achieved in 1983.
Notes: The Jetstream 31 is a derivative of the Handley Page H.P. 137 Jetstream, the original prototype of which was flown on 18 August 1967. The commuter version of the Jetstream was considered the prime configuration at the beginning of 1982 when all orders called for this version.

BAe JETSTREAM 31

Dimensions: Span, 52 ft 0 in (15,85 m); length, 47 ft 1½ in (14,36 m); height, 10 ft 6 in (3,20 m); wing area, 270 sq ft (25,08 m²).

BAe NIMROD MR Mk 2

Country of Origin: United Kingdom.

Type: Long-range maritime patrol aircraft.

Power Plant: Four 12,160 lb st (5 515 kgp) Rolls-Royce RB. 168-20 Spey Mk 250 turbofans.

Performance: Max. speed, 575 mph (926 km/h); max. transit speed, 547 mph (880 km/h); econ. transit speed, 490 mph (787 km/h); typical ferry range, 5,180–5,755 mls (8 340–9 265 km); typical endurance, 12 hrs.

Weights: Max. take-off, 177,500 lb (80 510 kg); max. overload, 192,000 lb (87 090 kg).

Armament: Ventral weapons bay accommodating full range of ASW weapons (e.g. Stingray homing torpedoes, mines, depth charges). Provision for two underwing pylons on each side for total of four Aérospatiale AS 12 missiles.

Accommodation: Normal operating crew of 12 with two pilots and flight engineer on flight deck and nine sensor operators and navigators in tactical compartment.

Status: Thirty-two Nimrod MR Mk 1s are being progressively brought up to MR Mk 2 standard in a programme continuing until mid-1984, the first MR Mk 2 having been accepted by the RAF on 23 August 1979. The first of 38 Nimrod MR Mk 1s was flown on 28 June 1968, a further eight being ordered in 1973. All remaining MR Mk 1s are being brought up to MR Mk 2 or modified to AEW Mk 3 (which see) standards.

Notes: The Nimrod MR Mk 2 possesses 60 times more computer power than the MR Mk 1 that it supplants, and is equipped with the advanced Searchwater maritime radar, an AQS-901 acoustics system compatible with the Barra sonobuoy, and EWSM (Electronic Warfare Support Measures) wingtip pods (as seen on drawing).

BAe NIMROD MR Mk 2

Dimensions: Span, 114 ft 10 in (35,00 m); length, 126 ft 9 in (38,63 m); height, 29 ft 8½ in (9,01 m); wing area, 2,121 sq ft (197,05 m²).

BAe NIMROD AEW MK 3

Country of Origin: United Kingdom.
Type: Airborne warning and control system aircraft.
Power Plant: Four 12,160 lb st (5 515 kgp) Rolls-Royce RB.
168-20 Spey Mk 250 turbofans.
Performance: No details have been released for publication,
but maximum and transit speeds are likely to be generally similar
to those of the MR Mk 2, and maximum endurance is in excess
of 10 hrs. The mission requirement calls for 6-7 hours on station
at 29,000–35,000 ft (8 840–10 670 m) at approx. 350 mph
(563 km/h) at 750–1,000 mls (1 120–1 600 km) from base.
Weights: No details available.
Accommodation: Flight crew of four and tactical team of six.
Tactical team comprises tactical air control officer, communications
control officer, EWSM (Electronic Warfare Support
Measures) operator and three air direction officers.
Status: Total of 11 Nimrod MR Mk 1 airframes being rebuilt to
AEW Mk 3 standard of which fully representative prototype flew
on 16 July 1980, and three aircraft had flown by the beginning
of 1982. Initial operational capability with the RAF is scheduled
for late 1982.
Notes: The Nimrod AEW Mk 3 is equipped with Marconi mission
system avionics with identical radar aerials mounted in nose
and tail, these being synchronised and each sequentially sweeping
through 180 deg in azimuth and providing uninterrupted
coverage through 360 deg of combined sweep. EWSM pods are
located at the wingtips and weather radar in the starboard wing
pinion tank.

BAe NIMROD AEW MK 3

Dimensions: Span, 115 ft 1 in (35,08 m); length, 137 ft 8½ in (41,97 m); height, 35 ft 0 in (10,67 m); wing area, 2,121 sq ft (197,05 m²).

BAe SEA HARRIER FRS Mk 1

Country of Origin: United Kingdom.

Type: Single-seat V/STOL shipboard multi-role fighter.

Power Plant: One 21,500 lb st (9 760 kgp) Rolls-Royce Pegasus 104 vectored-thrust turbofan.

Performance: (Estimated) Max. speed (clean aircraft), 595 mph (956 km/h) or Mach 0·9 at 36,000 ft (10 970 m), (with two Martel ASMs and two Sidewinder AAMs), 555 mph (892 km/h) or Mach 0·84 at 36,000 ft (10 970 m), 610 mph (980 km/h) or Mach 0·8 at sea level; tactical radius (intercept mission with two 30-mm cannon, two AAMs and two drop tanks), 500 mls (805 km); (strike mission with HI-LO-HI profile, 1,500 lb/680 kg external ordnance and two drop tanks), 450 mls (725 km); ferry range, 2,070 mls (3 330 km).

Weights: Operational empty, 12,500 lb (5 670 kg); max. loaded for short take-off, 22,500 lb (10 206 kg); max. take-off, 25,000 lb (11 339 kg).

Armament: Provision for two (flush-fitting) podded 30-mm Aden cannon beneath fuselage. Five external hardpoints (one fuselage and four wing) each stressed for 1,000 lb (453,5 kg), with max. external ordnance load for STO (excluding cannon) of 5,000 lb (2 268 kg). Typical loads include two Martel or Harpoon ASMs in inboard wing pylons and two AIM-9 Sidewinder AAMs on outboard pylons.

Status: First Sea Harrier (built on production tooling) flown on 21 August 1978, with deliveries against 34 ordered for Royal Navy commencing in second half of 1979, and some 20 delivered by the beginning of 1982. Six Sea Harrier FRS Mk 51s have been ordered by the Indian Navy for delivery from December 1982.

BAe SEA HARRIER FRS MK 1

Dimensions: Span, 25 ft 3 in (7,70 m); length, 47 ft 7 in (14,50 m); height, 12 ft 2 in (3,70 m); wing area, 201·1 sq ft (18,68 m²).

BEECHCRAFT COMMUTER C99

Country of Origin: USA.

Type: Short-haul regional transport.

Power Plant: Two (flat-rated) 715 shp Pratt & Whitney (Canada) PT6A-36 turboprops.

Performance: Max. speed, 308 mph (496 km/h); cruise (at max. cruise power), 285 mph (459 km/h) at 8,000 ft (2 440 m), 280 mph (450 km/h) at 16,000 ft (4 875 m); initial climb, 2,221 ft/min (11,3 m/sec); service ceiling, 28,080 ft (8 560 m); max. range (with 45 min reserves), 953 mls (1 534 km) at max. cruise power, 1,050 mls (1 690 km) at max. range power.

Weights: Empty, 6,124 lb (2 779 kg); max. take-off, 11,300 lb (5 126 kg).

Accommodation: Flight crew of two and 15 passengers in individual seats. Various alternative arrangements for mixed passenger/freight or all-freight operation.

Status: The prototype Commuter C99 (a conversion of a 1969 Model 99) was first flown on 20 June 1980, and the initial customer delivery was effected in August 1981. Production rate of two–three monthly at beginning of 1982.

Notes: The Commuter C99 is dimensionally similar to the original Model 99 (164 of which were produced between 1968 and 1975) but embodies over 700 design changes, among the most significant being those associated with higher operating weights, the PT6A-36 turboprops (which replace the lower-rated PT6A-27s powering the B99 when production ended), a new wing mainspar, hydraulic undercarriage actuation and a simplified electrical system based on that of the King Air.

BEECHCRAFT COMMUTER C99

Dimensions: Span, 45 ft 10½ in (13,99 m); length, 44 ft 6¾ in (13,58 m); height, 14 ft 4¼ in (4,38 m); wing area, 279·7 sq ft (25,98 m²).

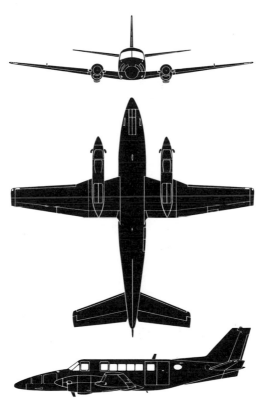

BEECHCRAFT COMMUTER 1900

Country of Origin: USA.

Type: Short-haul regional transport.

Power Plant: Two (flat-rated) 1,000 shp Pratt & Whitney (Canada) PT6A-65B turboprops.

Performance: (Estimated) Max. cruising speed, 303 mph (488 km/h) at 10,000 ft (3 050 m); initial climb, 2,280 ft/min (11,6 m/sec); service ceiling, 30,000 ft (9 150 m); range (max. cruise power and full load), 639 mls (1 028 km) at 10,000 ft (3 050 m), 977 mls (1 572 km) at 25,000 ft (7 620 m).

Weights: Empty, 8,500 lb (3 856 kg); max. take-off, 15,245 lb (6 915 kg).

Accommodation: Flight crew of two and up to 19 passengers in individual seats.

Status: Flight testing of first of three pre-series examples scheduled to commence during spring 1982, with initial customer deliveries in 1983.

Notes: The Commuter 1900 is essentially a stretched-fuselage derivative of the Super King Air designed specifically for the regional transport market and for multi-stop operations without refuelling. The fuselage possesses adequate width for single seats to be track-mounted each side of a central aisle, and two tons of cargo, or mixed passenger/freight loads may be carried, an enlarged rear cargo-loading door being an option. A corporate executive transport version of the Commuter 1900, the Model 1200, was under consideration at the beginning of 1982. The 19-passenger Commuter 1900 complements the 15-passenger Commuter C99 (see pages 46–47) in the Beechcraft range.

BEECHCRAFT COMMUTER 1900

Dimensions: Span, 54 ft 6 in (16,61 m); length, 57 ft 10 in (17,63 m); height, 14 ft 10¾ in (4,53 m); wing area, 303 sq ft (28,15 m²).

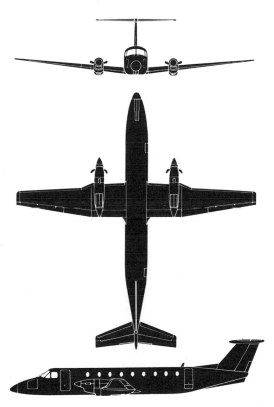

BEECHCRAFT T-34C (TURBINE MENTOR)

Country of Origin: USA.

Type: Tandem two-seat primary trainer.

Power Plant: One (flat-rated) 400 shp Pratt & Whitney (Canada) PT6A-25 turboprop.

Performance: Max. speed, 246 mph (396 km/h) at 17,000 ft (5 180 m); cruising speed, 208 mph (335 km/h) at 1,000 ft (305 m), 233 mph (375 km/h) at 10,000 ft (3 050 m); initial climb, 1,480 ft/min (7,52 m/sec); time to 20,000 ft (6 095 m), 15 min; range, 492 mls (792 km) at 1,000 ft (305 m), 602 mls (970 km) at 10,000 ft (3 050 m), 815 mls (1 312 km) at 20,000 ft (6 095 m).

Weights: Empty, 2,940 lb (1 334 kg); max. take-off, 4,300 lb (1 950 kg).

Status: First of two YT-34Cs flown 21 September 1973, and first T-34C for US Navy in August 1976. Total of 196 ordered by beginning of 1982 against total US Navy requirement for 278.

Notes: An export version of the T-34C with a PT6A-25 flat-rated at 550 shp, wing racks for external ordnance (two 600-lb/ 272-kg capacity inboard and two 300-lb/136-kg capacity stations outboard) and an armament control system is designated T-34C-1. This version has been exported to Argentina (15), Ecuador (23), Gabon (4), Indonesia (16), Morocco (12), Peru (6) and Uruguay (3). Six have also been supplied to Algeria without armament provisions as Turbine Mentor 34Cs and serve with the national pilot training school. The T-34C-1 is suitable for both armament training and light counter-insurgency or close air support missions.

BEECHCRAFT T-34C (TURBINE MENTOR)

Dimensions: Span, 33 ft 4¾ in (10,18 m); length, 28 ft 8½ in (8,75 m); height, 9 ft 10⅞ in (3,02 m); wing area, 179·56 sq ft (16,68 m²).

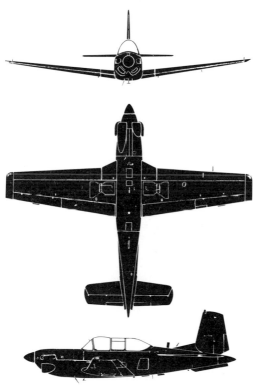

BOEING 727-200

Country of Origin: USA.

Type: Short/medium-haul commercial transport.

Power Plant: Three 14,500 lb st (6 577 kgp) Pratt & Whitney JT8D-9, 15,000 lb st (6 805 kgp) JT8D-11 or 15,500 lb st (7 030 kgp) JT8D-15 turbofans.

Performance: Max. speed, 621 mph (999 km/h) at 20,500 ft (6 250 m); max. cruise, 599 mph (964 km/h) at 24,700 ft (7 530 m); econ. cruise, 570 mph (917 km/h) at 30,000 ft (9 145 m); range (with max. payload), 1,845 mls (2 970 km), (with max. fuel), 2,850 mls (4 585 km).

Weights: Operational empty, 97,525 lb (44 235 kg); max. take-off, 208,000 lb (94 347 kg).

Accommodation: Crew of three on flight deck and basic arrangement for 163 passengers in six-abreast seating, with max. seating for 189 passengers.

Status: First Model 727-100 flown 9 February 1963, with first delivery (to United) following 29 October 1963. Stretched 727-200 flown 27 July 1967, with first delivery (to Northeast) following 11 December 1967. Sales totalled 1,824 aircraft by beginning of 1982, with 1,785 delivered, and production was then tapering off with planned deliveries of 26 during course of year.

Notes: Claiming the distinction of being the world's biggest selling jet airliner, the Model 727 is expected to be phased out of production in 1983–84, but in 1982 consideration is being given to the possibility of re-engining existing aircraft with the new Pratt & Whitney PW2037 turbofan. This project (which applies only to -200 series aircraft) is referred to as the Model 727RE and envisages the confinement of structural modifications to the rear fuselage, the elimination of the centre engine dictating new vertical tail surfaces. An alternative engine under consideration for the Model 727RE is the Rolls-Royce RB.211-535.

BOEING 727-200

Dimensions: Span, 108 ft 0 in (32,92 m); length, 153 ft 2 in (46,69 m); height, 34 ft 0 in (10,36 m); wing area, 1,560 sq ft (144,92 m²).

BOEING 737-200

Country of Origin: USA.

Type: Short-haul commercial transport.

Power Plant: Two 16,000 lb st (7 258 kgp) Pratt & Whitney JT8D-17 turbofans.

Performance: Max. cruising speed, 564 mph (908 km/h) at 25,000 ft (7 620 m); econ. cruise, 502 mph (808 km/h) at 33,000 ft (10 055 m); long-range cruise, 481 mph (775 km/h) at 35,000 ft (10 670 m); range (with max. fuel), 3,086 mls (4 967 km), (with max. payload), 1,750 mls (2 817 km).

Weights: Operational empty, 61,210 lb (27 764 kg); Max. take-off, 117,000 lb (53 071 kg).

Accommodation: Flight crew of two and up to 130 passengers in six-abreast seating, with optional arrangement for 115 passengers.

Status: First Model 737 flown 9 April 1967, with first deliveries (to Lufthansa) following same year. Stretched 737-200 flown on 8 August 1967. A total of 978 ordered by beginning of 1982, with delivery of 113 scheduled during course of year.

Notes: On 26 March 1981, Boeing announced its intention to proceed with a further stretched version of the basic Model 737. This, the 737-300 due to fly in April 1984 with deliveries commencing in the following December, will have 20,000 lb st (9 072 kgp) CFM 56-3 turbofans and an 8 ft 8 in (2,64 m) fuselage stretch to provide for 17–20 extra seats. The wing is to be strengthened, and the leading- and trailing-edge high-lift devices and tail surfaces modified, and standard and optional maximum take-off weights will be 124,500 lb (56 473 kg) and 130,000 lb (58 968 kg).

BOEING 737-200

Dimensions: Span, 93 ft 0 in (28,35 m); length, 100 ft 0 in (30,48 m); height, 37 ft 0 in (11,28 m); wing area, 980 sq ft (91,05 m²).

BOEING 747-200B (EUD)

Country of Origin: USA.

Type: Long-haul commercial transport.

Power Plant: Four 54,750 lb st (24 835 kgp) Pratt & Whitney JT9D-7R4G2 turbofans.

Performance: Max. cruising speed, 583 mph (939 km/h) at 35,000 ft (10 670 m); econ. cruise, 564 mph (907 km/h) at 35,000 ft (10 670 m); long-range cruise, 558 mph (898 km/h); range (max. payload at econ. cruise), 6,860 mls (11 040 km), (max. fuel at long-range cruise), 8,606 mls (13 850 km).

Weights: Operational empty, 390,130 lb (176 962 kg); max. take-off, 833,000 lb (377 850 kg).

Accommodation: Normal flight crew of three and up to 69 passengers six-abreast on extended upper deck (EUD) plus basic accommodation for 410 passengers in mixed-class arrangement, or 415 nine-abreast or 484 10-abreast in economy class seating.

Status: First EUD Model 747-200B (which will also be first with the new generation JT9D-7R4 engines) is scheduled to fly October 1982, with delivery (to Swissair) in March 1983. First Model 747-100 flown 9 February 1969, and first -200 on 11 October 1970. Orders (all versions) totalled 586 at beginning of 1982, with 28 scheduled to be delivered during year.

Notes: The extended upper deck is available to order on all new-production Model 747s (apart from the 747SP), the bulged upper fuselage being stretched aft by 23 ft (7,01 m) to increase the deck's economy class seating from 32 to 69 passengers. Max. take-off weight remains unchanged but operational empty weight is increased by 8,000 lb (3 630 kg).

BOEING 747-200B (EUD)

Dimensions: Span, 195 ft 8 in (59,64 m); length, 231 ft 4 in (70,51 m); height, 63 ft 5 in (19,33 m); wing area, 5,685 sq ft (528,15 m²).

BOEING 757-200

Country of Origin: USA.

Type: Short/medium-haul commercial transport.

Power Plant: Two 37,300 lb st (16 920 kgp) Rolls-Royce RB.211-535C turbofans.

Performance: Max. cruising speed, 568 mph (915 km/h) at 29,000 ft (8 840 m); long-range cruise, 528 mph (850 km/h) at 37,000 ft (11 280 ft); range (max. payload at econ. cruise), 1,380 mls (2 220 km), (max. fuel at long-range cruise), 2,895 mls (4 660 km).

Weights: Operational empty, 130,670 lb (59 272 kg); max. take-off, 220,000 lb (99 790 kg).

Accommodation: Flight crew of two (with provision for third member) and typical accommodation for 178 mixed class or 196 tourist class passengers, with max. capacity of 233 passengers.

Status: First Model 757 was scheduled to fly in February 1982, with first two aircraft (to Eastern) expected to be delivered in December 1982. Firm orders totalled 136 aircraft at beginning of 1982.

Notes: Two versions of the Model 757-200—the current standard-body aircraft—are proposed, these being the physically similar Basic and Medium Range aircraft, the latter being certificated at 230,000 lb (104 328 kg). An increased gross weight version with a slightly heavier structure and a take-off weight of 240,000 lb (108 864 kg) is under consideration and Pratt & Whitney PW2037 turbofans are being offered as options to the RB.211s for aircraft delivered from late 1984. The Model 757 is the first Boeing airliner to be launched with a non-US engine.

BOEING 757-200

Dimensions: Span, 124 ft 6 in (37,82 m); length, 155 ft 3 in (47,47 m); height, 44 ft 6 in (13,56 m); wing area, 1,951 sq ft (181,25 m²).

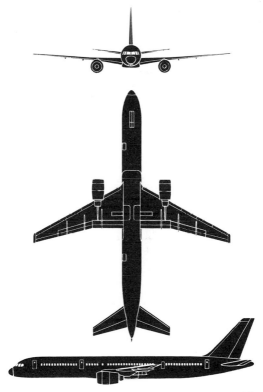

BOEING 767-200

Country of Origin: USA.

Type: Medium-haul commercial transport.

Power Plant: Two 47,700 lb st (21 637 kgp) Pratt & Whitney JT9D-7R4D or 47,900 lb st (21 727 kgp) General Electric CF6–80A turbofans.

Performance: (CF6-80A engines) Max. cruising speed, 579 mph (932 km/h) at 30,000 ft (9 145 m); econ. and long-range cruise, 528 mph (850 km/h) at 39,000 ft (11 885 m); range (with max. payload), 2,554 mls (4 110 km), (max. fuel), 5,748 mls (9 250 km).

Weights: Operational empty 180,210 lb (81 743 kg); max. take-off, 300,000 lb (136 080 kg).

Accommodation: Flight crew of two plus optional third crew member on flight deck. Typical mixed-class seating for 18 six-abreast and 193 seven-abreast, with max. single-class seating for 255 passengers seven-abreast.

Status: First Model 767 flown on 26 September 1981, and three additional aircraft had flown by the beginning of 1982, when orders totalled 173 aircraft (with 138 more on option). First customer delivery (to United Airlines) scheduled for August 1982 delivery, and 28 Model 767s to be delivered during course of year and 80 during 1983, with production rate of eight per month planned for following year.

Notes: A stretched version of the Model 767 with a 29 ft 4 in (8,94 m) lengthening of the fuselage and 285-passenger capacity was under consideration at the beginning of 1982 for introduction in 1985–86.

BOEING 767-200

Dimensions: Span, 156 ft 4 in (47,65 m); length, 159 ft 2 in (48,50 m); height, 52 ft 0 in (15,85 m); wing area, 3,050 sq ft (283,3 m²).

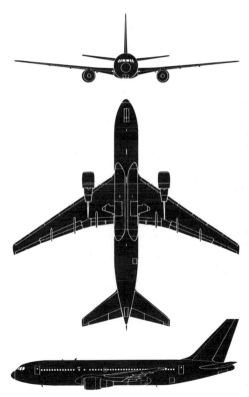

BOEING E-3A SENTRY

Country of Origin: USA.

Type: Airborne warning and control system aircraft.

Plant Plant: Four 21,000 lb st (9 525 kgp) Pratt & Whitney TF33-PW-100A turbofans.

Performance: (At max. weight) Average cruising speed, 479 mph (771 km/h) at 28,900–40,100 ft (8 810–12 220 m); average loiter speed, 376 mph (605 km/h) at 29,000 ft (8 840 m); time on station (unrefuelled) at 1,150 mls (1 850 km) from base, 6 hrs, (with one refuelling), 14·4 hrs; ferry range, 5,034 mls (8 100 km) at 475 mph (764 km/h).

Weights: Empty, 170,277 lb (77 238 kg); normal loaded, 214,300 lb (97 206 kg); max. take-off, 325,000 lb (147 420 kg).

Accommodation: Operational crew of 17 comprising flight crew of four, systems maintenance team of four, a battle commander and an air defence operations team of eight.

Status: First of two (EC-137D) development aircraft flown 9 February 1972, two pre-production E-3As following in 1975. Total of 31 Sentries in process of delivery to the USAF with final aircraft scheduled to be received in May 1984. Eighteen similar aircraft being supplied to NATO (excluding UK), with first airframe delivered to Dornier (for mission avionics installation) on 31 March 1981, and last to be delivered June 1985. A further five Sentries are to be supplied to Saudi Arabia from August 1985.

Notes: Full-scale development of upgraded communications, command and control equipment for retrofit to the USAF's first 24 E-3As was initiated mid-1981.

BOEING E-3A SENTRY

Dimensions: Span, 145 ft 9 in (44,42 m); length, 152 ft 11 in (46,61 m); height, 42 ft 5 in (12,93 m); wing area, 2,892 sq ft (268,67 m²).

CANADAIR CL-601 CHALLENGER

Country of Origin: Canada.

Type: Light business executive transport.

Power Plant: Two 8,650 lb st (3 924 kgp) General Electric CF34-1A turbofans.

Performance: Max. cruising speed, 518 mph (834 km/h) or Mach 0·78 at 40,000 ft (12 190 m); normal cruise, 497 mph (800 km/h) or Mach 0·75; range cruise, 460 mph (740 km/h) or Mach 0·7; range (with 3,925 lb/1 780 kg payload and reserves), 3,685 mls (5 930 km) at Mach 0·78, 4,090 mls (6 580 km) at Mach 0·7.

Weights: Operational empty, 24,075 lb (10 920 kg); max. take-off, 41,650 lb (18 892 kg).

Accommodation: Flight crew of two with typical arrangements in main cabin for 8–11 passengers in executive configuration. Optional arrangements for 18 passengers and 28 passengers for low- and high-density commuterliner operation, and mixed passenger/freight configuration.

Status: First CL-601 version of Challenger scheduled to fly April 1982, with certification one year later. First CL-600 flown on 8 November 1978, with customer deliveries commencing 1981. Firm orders for Challengers at beginning of 1982 totalled 138 of which 11 were for CL-601 version.

Notes: The CL-601 is similar to the CL-600 (see 1981 edition) apart from engines, the former having 7,500 lb st (3 405 kgp) ALF 502L-2 turbofans. The postponement of development of the "stretched" CL-610 was announced in August 1981 after design reassessment.

CANADAIR CL-601 CHALLENGER

Dimensions: Span, 61 ft 10 in (18,85 m); length, 68 ft 5 in (20,85 m); height, 20 ft 8 in (6,30 m); wing area, 450 sq ft (41,82 m²).

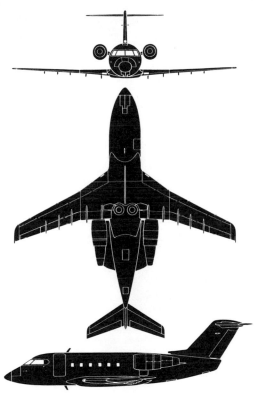

CAPRONI VIZZOLA C 22J

Country of Origin: Italy.
Type: Side-by-side two-seat primary and continuation trainer and light reconnaissance aircraft.
Power Plant: Two 202 lb st (92 kgp) Microturbo TRS 18-046 turbojets, or (production option) two 242 lb st (110 kgp) Klockner-Humboldt-Deutz KHD-317 turbojets.
Performance: (Estimated with KHD-317 engines) Max. speed, 329 mph (530 km/h) at 8,200 ft (2 500 m); max. continuous cruise, 292 mph (470 km/h) at 16,405 ft (5 000 m); econ. cruise, 186 mph (300 km/h) at 9,840 ft (3 000 m); initial climb, 2,070 ft/min (10,51 m/sec); time to 16,405 ft (5 000 m), 12 min; range (internal fuel with 10% reserves), 660 mls (1 060 km).
Weights: Empty, 1,124 lb (510 kg); normal loaded, 1,984 lb (900 kg); max. take-off, 2,425 lb (1 100 kg).
Armament: (Weapons training role) Two or four standard NATO wing pylons for which typical loads can include four 97-lb (44-kg) or 440-lb (200-kg) practice bombs, two 7,62-mm gun pods with 500 rounds, or two pods each containing 18 2-in (5-cm) rockets.
Status: Prototype C 22J flown 21 July 1980, and development continuing (under Agusta aegis) at beginning of 1982 as a low-cost continuation trainer and light surveillance/reconnaissance aircraft (C 22R). Pre-series of four C 22Js under construction, with initial series of 20 to be launched during 1982.
Notes: Offering low initial procurement and minimum operational cost, the C 22J owes much to the manufacturer's experience in the design and development of advanced sailplanes, the fuselage, utilising a fibreglass shell, acting as a lifting body.

CAPRONI VIZZOLA C 22J

Dimensions: Span, 32 ft 9½ in (10,00 m); length, 20 ft 3¼ in (6,19 m); height, 6 ft 2 in (1,88 m); wing area, 94·19 sq ft (8,75 m²).

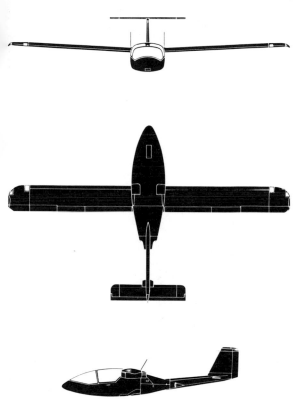

CASA C-101 AVIOJET

Country of Origin: Spain.
Type: Tandem two-seat basic and advanced trainer.
Power Plant: One 3,500 lb st (1 588 kgp) Garrett AiResearch TFE 731-2-2J turbofan.
Performance: Max. speed, 479 mph (770 km/h) or Mach 0·7 at 28,000 ft (8 535 m), 404 mph (650 km/h) or Mach 0·53 at sea level; time to 25,000 ft (7 620 m), 12 min; service ceiling, 41,000 ft (12 495 m); range (max. internal fuel), 2,485 mls (4 000 km).
Weights: Operational empty, 6,790 lb (3 080 kg); normal loaded, 10,362 lb (4 700 kg); max. take-off, 12,346 lb (5 600 kg).
Armament: (C-101 BB) Seven external stores stations for max. of 3,307 lb (1 500 kg) of ordnance.
Status: First of four prototypes flown 29 June 1977. Deliveries of production C-101 EB to Spanish Air Force commenced March 1980, and some 45 delivered to the service by beginning of 1982 against orders for 88 aircraft. First export C-101 BB delivered August 1981 against initial order for eight from Chile. Production rate of three monthly.
Notes: The export C-101 BB differs from the C-101 EB for the Spanish Air Force primarily in having a 3,700 lb st (1 678 kgp) TFE 731-3-1J turbofan and provision for interchangeable ventral packs containing either one 30-mm DEFA 553 cannon or twin 12,7-mm M3 machine guns, six underwing stores pylons, an optical weapons sight and an armament control system. It is anticipated that licence assembly of the C-101 BB will be undertaken in Chile, the Chilean Air Force having a requirement for 60–70 aircraft in the Aviojet category.

CASA C-101 AVIOJET

Dimensions: Span, 34 ft 9⅜ in (10,60 m); length, 40 ft 2¼ in (12,25 m); height, 13 ft 11 in (4,25 m); wing area, 215·3 sq ft (20,00 m²).

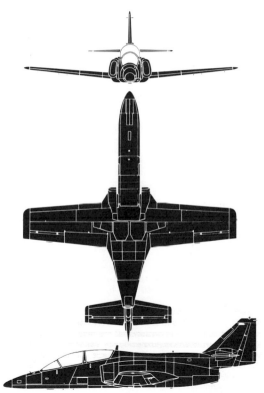

CASA C-212 AVIOCAR SERIES 200

Country of Origin: Spain.
Type: STOL utility and regional commercial transport.
Power Plant: Two 900 shp (944 eshp) Garrett AiResearch TPE 331-10-501-C turboprops.
Performance: Max. cruising speed, 227 mph (365 km/h) at 10,000 ft (3 050 m); econ. cruise, 216 mph (347 km/h) at 10,000 ft (3 050 m); max. initial climb, 1,700 ft/min (8,64 m/sec); service ceiling, 28,000 ft (8 535 m); Range (26 passengers and 45 min reserves), 230 mls (370 km), (16 passengers and same reserves), 840 mls (1 352 km) at normal cruise, 933 mls (1 500 km) at long-range cruise.
Weights: Empty equipped, 9,072 lb (4 115 kg); max. take-off, 16,424 lb (7 450 kg).
Accommodation: Flight crew of two and (regional airline arrangement) 26 passengers in four-abreast seating with central aisle or 19 passengers in three-abreast seating.
Status: First prototype C-212 flown 26 March 1971, the 138th and 139th production aircraft serving as prototypes for the Series 200, the first of these flying on 30 April 1978. Deliveries of Series 200 commenced early 1980, and total sales (Series 100 and 200) were approximately 300 at the beginning of 1982, with more than 200 delivered and production running at four monthly.
Notes: The C-212 is available in several civil and military versions, the latter including navigational training, photographic survey, and ASW and maritime surveillance variants, all of which have been supplied to the Spanish Air Force. An assembly line for the C-212 is operated in Indonesia by Nurtanio, 29 Series 100 Aviocars having been produced by this company before assembly and part-manufacture switched to the Series 200.

CASA C-212 AVIOCAR SERIES 200

Dimensions: Span, 62 ft 4 in (19,00 m); length, 49 ft 10½ in (15,20 m); height, 20 ft 8¾ in (6,32 m); wing area, 430·56 sq ft (40,00 m²).

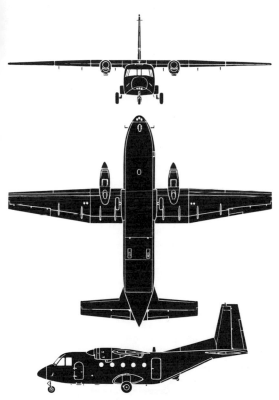

CESSNA MODEL 303 CRUSADER

Country of Origin: USA.

Type: Light cabin monoplane.

Power Plant: Two 250 bhp Teledyne Continental TSIO-520-AE turbo-supercharged six-cylinder horizontally-opposed engines.

Performance: Max. speed, 249 mph (400 km/h) at 18,000 ft (5 485 m); cruise (71% power), 225 mph (363 km/h) at 20,000 ft (6 095 m), (72% power), 207 mph (333 km/h) at 10,000 ft (3 050 m); initial climb, 1,480 ft/min (7,52 m/sec); service ceiling, 25,000 ft (7 620 m); max. range, 1,156 mls (1 861 km) at 20,000 ft (6 095 m), 1,174 mls (1 889 km) at 10,000 ft (3 050 m).

Weights: Empty, 3,305 lb (1 499 kg); max. take-off, 5,150 lb (2 336 kg).

Accommodation: Six individual seats in three pairs with 10-in (25-cm) aisle and baggage lockers in nose, wing and aft cabin for total of 590 lb (268 kg).

Status: First of two prototypes flown on 17 October 1979, with type certification obtained on 24 August 1981, and initial customer deliveries commencing in the following month. A total of 280 Crusaders is scheduled for delivery during the 1982 Fiscal Year.

Notes: Originally evolved from a mid 'seventies design concept for a twin-engined trainer, the Crusader is the first entirely new 'twin' to be marketed by Cessna for 10 years.

CESSNA MODEL 303 CRUSADER

Dimensions: Span, 38 ft 10 in (11,84 m); length, 30 ft 5 in (9,27 m); height, 13 ft 4 in (4,06 m); wing area, 189·2 sq ft (17,60 m²).

CESSNA CITATION II

Country of Origin: USA.

Type: Light business executive transport.

Power Plant: Two 2,500 lb st (1 135 kgp) Pratt & Whitney (Canada) JT15D-4 turbofans.

Performance: Max. cruising speed, 420 mph (676 km/h) at 25,400 ft (7 740 m); long-range cruise, 380 mph (611 km/h) at 43,000 ft (13 105 m); range (eight passengers and 45 min reserves), 2,080 mls (3 347 km) at 380 mph (611 km/h); initial climb, 3,500 ft/min (17,8 m/sec); time to 41,000 ft (12 495 m), 34 min; max. cruise altitude, 43,000 ft (13 105 m).

Weights: Empty equipped (typical), 6,960 lb (3 160 kg); max. take-off, 12,500 lb (5 675 kg).

Accommodation: Normal flight crew of two on separate flight deck and various arrangements for up to 10 passengers in main cabin.

Status: First of two prototypes of the Citation II flown on 31 January 1977, with first customer deliveries commencing late March 1978, with some 330 delivered by the beginning of 1982, when combined production rate of Citation I and II was 15 monthly, some 280 Citation Is having been delivered. These were preceded by 349 of the original Citation.

Notes: The Citation II is a stretched (4 ft/1,22 m longer cabin) version of the original Citation, with a higher aspect ratio wing, uprated engines and increased fuel capacity. It is being manufactured in parallel with the Citation I and I/SP (the latter catering for single-pilot operation) which have similar accommodation to the first Citation (five–seven passengers), JT15D-1A turbofans and a 47 ft 1 in (14,36 m) wing span. Deliveries of the Citation I commenced in February 1977.

CESSNA CITATION II

Dimensions: Span, 51 ft 8 in (15,76 m); length, 47 ft 3 in (14,41 m); height, 14 ft 11 in (4,55 m).

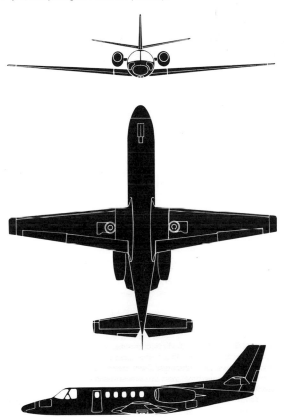

CESSNA CITATION III

Country of Origin: USA.
Type: Light business executive transport.
Power Plant: Two 3,650 lb st (1 656 kgp) Garrett AiResearch TFE 731-3B-100S turbofans.
Performance: Max. cruising speed, 540 mph (869 km/h) at 33,000 ft (10 060 m), 528 mph (850 km/h) at 41,000 ft (12 500 m); max. initial climb, 4,475 ft/min (22,73 m/sec); ceiling, 51,000 ft (15 545 m); range (six passengers and 45 min reserves), 2,858 mls (4 600 km).
Weights: Operational empty, 9,985 lb (4 529 kg); max. take-off, 19,700 lb (8 936 kg).
Accommodation: Normal flight crew of two on separate flight deck with standard cabin arrangement for six passengers in individual seats. Optional arrangements for eight to 13 passengers.
Status: First of two prototypes flown on 30 May 1979, with certification scheduled for April 1982. First production Citation III was scheduled for completion December 1981, and customer deliveries are planned to commence in December 1982, production rate being four monthly by June 1983, and increasing to 10 monthly by May 1985.
Notes: The Citation III possesses no commonality with the Citation II (see pages 74–75) despite its name. Various changes have been made to the wing trailing edge and the flap system during the course of 1981, and plans to offer an extended long-range version with an additional fuel cell aft of the cabin's rear pressure bulkhead have now been abandoned. Flight testing has indicated that the maximum operating speed of the series version will be increased from Mach 0·81 to 0·83, and that definitive performance figures for climb and level speed will be improved upon those quoted above.

CESSNA CITATION III

Dimensions: Span, 53 ft 3½ in (16,30 m); length, 55 ft 6 in (16,90 m); height, 17 ft 3½ in (5,30 m); wing area, 312 sq ft (29,00 m²).

DASSAULT-BREGUET ATLANTIC NG

Country of Origin: France.
Type: Long-range maritime patrol aircraft.
Power Plant: Two 5,665 shp Rolls-Royce/SNECMA Tyne RTy 20 Mk 21 turboprops.
Performance: Max. speed, 368 mph (593 km/h) at sea level; normal cruise, 345 mph (556 km/h) at 25,000 ft (7 620 m); typical patrol speed, 196 mph (315 km/h); initial climb, 2,000 ft/min (10,1 m/sec); service ceiling, 30,000 ft (9 100 m); typical mission, 8 hrs patrol at 196 mph/315 km/h at 2,000–3,000 ft (610–915 m) 690 mls (1 110 km) from base; max. endurance, 18 hrs.
Weights: Empty equipped, 55,115 lb (25 000 kg); normal loaded, 97,885 lb (44 400 kg); max. take-off, 101,850 lb (46 200 kg).
Armament: Up to eight Mk 46 homing torpedoes, nine 550-lb (250-kg) bombs or 12 depth charges, plus two AM 39 Exocet ASMs in forward fuselage weapons bay, plus up to 78 sonobuoys in rear bay. Four wing hardpoints with combined capacity of 7,715 lb (3 500 kg).
Status: First of two prototype Atlantic NGs (converted from Atlantic I No 42 airframe) flown on 8 May 1981, with second prototype (converted from Atlantic I No 69) scheduled to fly spring 1982. France's Aéronavale has a requirement for 42 Atlantic NGs with deliveries scheduled to commence 1985.
Notes: The Atlantic NG (*Nouvelle Génération*) is a modernised version of the Atlantic I, production of which terminated in 1973 after completion of 87 series aircraft.

DASSAULT-BREGUET ATLANTIC NG

Dimensions: Span, 122 ft 7 in (37,36 m); length, 107 ft 0¼ in (36,62 m); height, 37 ft 1¼ in (11,31 m); wing area, 1,295·3 sq ft (120,34 m²).

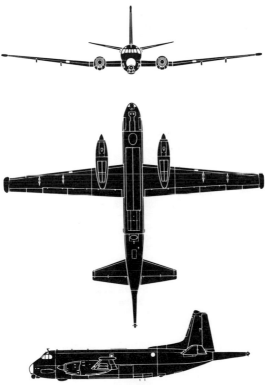

DASSAULT-BREGUET HU-25A GUARDIAN

Country of Origin: France.
Type: Light maritime surveillance aircraft.
Power Plant: Two 5,440 lb st (2 468 kgp) Garrett AiResearch ATF-3-6-2C turbofans.
Performance: Max. cruising speed, 531 mph (855 km/h) or Mach 0·8 at 40,000 ft (12 200 m); econ. cruise, 475 mph (765 km/h); max. range (with 5% reserves plus 30 min at sea level), 2,590 mls (4 170 km).
Weights: Empty, 19,000 lb (8 620 kg); max. take-off, 32,000 lb (14 515 kg).
Accommodation: Crew of five for US Coast Guard MRS (Medium Range Surveillance) mission, comprising two pilots, surveillance systems operator and two visual observers. Three passenger seats on port side of cabin. Crew of six for Aéronavale (Gardian) version, comprising two pilots, two visual observers, a navigator and a radar operator. Four passengers on four-seat couch and provision for additional four-seat couch.
Status: First fully representative prototype flown on 30 April 1980, with definitive prototype flown 15 April 1981. Five ordered by French Aéronavale (as Gardian) with deliveries scheduled from late 1982, and first of 41 for US Coast Guard (as HU-25A Guardian) was expected to be formally accepted early 1982. Deliveries of commercial equivalent, the Mystère-Falcon 200, scheduled for mid-1983, with projected delivery rate of two monthly.
Notes: The Guardian/Gardian is a maritime surveillance version of the Mystère-Falcon 200 (formerly 20H) business executive transport. Approximately 480 Series 20/200 aircraft had been sold by the beginning of 1982 with more than 450 delivered.

DASSAULT-BREGUET HU-25A GUARDIAN

Dimensions: Span, 53 ft 6 in (16,30 m); length, 56 ft 3 in (17,15 m); height, 17 ft 5 in (5,32 m); wing area, 450 sq ft (41,80 m²).

DASSAULT-BREGUET MIRAGE F1

Country of Origin: France.

Type: Single-seat multi-role fighter.

Power Plant: One 11,023 lb st (5 000 kgp) dry and 15,873 lb st (7 200 kgp) reheat SNECMA Atar 9K50 turbojet.

Performance: (F1C) Max. speed (clean aircraft), 914 mph or Mach 1·2 at sea level, 1,450 mph (2 355 km/h) or Mach 2·2 at 39,370 ft (12 000 m); cruise, 550 mph (885 km/h) at 29,530 ft (9 000 m); initial climb, 41.930 ft/min (213 m/sec); service ceiling, 65,600 ft (20 000 m); radius (max. external ordnance and HI-LO-HI mission profile), 260 mls (418 km), (with two drop tanks and 4,410 lb/2 000 kg bomb load), 670 mls (1 078 km); ferry range, 2,050 mls (3 300 km).

Weights: Empty, 16,314 lb (7 400 kg); normal loaded, 24.030 lb (10 900 kg); max. take-off, 32,850 lb (14 900 kg).

Armament: Two 30-mm DEFA 553 cannon and (intercept mission) one–three Matra 550 Magic plus two AIM-9 Sidewinder AAMs, or (close support) up to 8,818 lb (4 000 kg) of ordance.

Status: First of four prototypes flown 23 December 1966, and first production aircraft flown 15 February 1973. Production continuing at rate of seven monthly at beginning of 1982 against contracts for 253 for Armée de l'Air (80 F1Cs, 120 F1C-200s, 21 F1CRs and 14 two-seat F1Bs) and some 400 for export to 10 customers, continued production being assured until the end of 1984.

Notes: Current production models for the Armée de l'Air consist of the F1C-200 with a fixed flight refuelling probe, the tactical reconnaissance F1CR and the two-seat F1B conversion trainer. Whereas the F1C is an air–air version, the export F1A and F1E are optimised for the air–ground role, the former for VFR operations only. A Libyan Mirage F1ED is illustrated above.

DASSAULT-BREGUET MIRAGE F1

Dimensions: Span, 27 ft 6¾ in (8,40 m); length, 49 ft 2½ in (15,00 m); height, 14 ft 9 in (4,50 m); wing area, 269·1 sq ft (25,00 m²).

DASSAULT-BREGUET MIRAGE 2000

Country of Origin: France.

Type: Single-seat multi-role fighter.

Power Plant: One 12,345 lb st (5 600 kgp) dry and 19,840 lb st (9 000 kgp) reheat SNECMA M53-5 turbofan.

Performance: Max. speed (clean aircraft), 915 mph (1 472 km/h) or Mach 1·2 at sea level, 1,550 mph (2 495 km/h) or Mach 2·35 above 36,090 ft (11 000 m); max. climb, 49,000 ft/min (249 m/sec); service ceiling, 65,000 ft (19 800 m); time to 49,200 ft (15 000 m) and Mach 2·0, 4·0 min; combat radius (intercept mission with two drop tanks and four AAMs), 435 mls (700 km); ferry range, 2,420 mls (3 900 km).

Weights: Combat loaded, 19,840 lb (9 000 kg); max. take-off 33,070 lb (15 000 kg).

Armament: Two 30-mm DEFA 554 cannon and (air superiority) two Matra 550 Magic and two Matra Super 530D AAMs, or (close support) up to 13,227 lb (6 000 kg) of ordnance.

Status: First of five prototypes flown 10 March 1978, with last (two-seat Mirage 2000B) on 11 October 1980. First of two two-seat low-level Mirage 2000N (Nucléaire) scheduled to enter flight test in 1983. Total of 73 (including Fiscal 1982 procurement) ordered for Armée de l'Air against anticipated requirement for 400 in three main versions. India negotiating procurement of 150 (including licence assembly and manufacture) early 1982.

Notes: To be procured by the Armée de l'Air in single-seat air superiority and attack, and two-seat conversion training and low-level penetration versions, the Mirage 2000 is scheduled to enter service from 1984, the M53-5 turbofan giving place to the M53-P from 1985, this having military and reheat ratings of 14,330 lb st (6 500 kgp) and 21,385 lb st (9 700 kgp) respectively.

DASSAULT-BREGUET MIRAGE 2000

Dimensions: Span, 29 ft 6⅓ in (9,000 m); length, 50 ft 3½ in (15,33 m); wing area, 441·3 sq ft (41,00 m²)

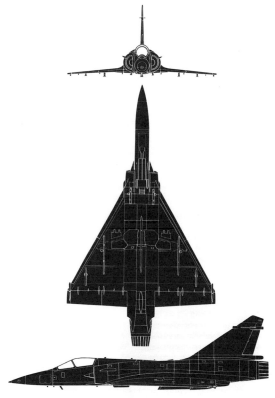

DASSAULT-BREGUET
MYSTÈRE-FALCON 50

Country of Origin: France.
Type: Light business executive transport.
Power Plant: Three 3,700 lb st (1 680 kgp) Garrett AiResearch TFE731-3 turbofans.
Performance: Max. cruising speed, 547 mph (880 km/h) or Mach 0·82 at 33,000 ft (10 060 m); long-range cruise, 495 mph (792 km/h) at 37,000 ft (11 275 m); range (eight passengers and 45 min reserves), 4,088 mls (6 578 km) at long-range cruise, 2,764 mls (4 447 km) at max. cruise.
Weights: Empty equipped, 19,840 lb (9 000 kg); max. take-off, 38,800 lb (17 600 kg).
Accommodation: Flight crew of two and various arrangements in main cabin for six to twelve passengers.
Status: First prototype flown on 7 November 1976, with first pre-series aircraft following 13 June 1978. Approximately 180 ordered by beginning of 1982, when production rate was 4·5 monthly with some 75 delivered.
Notes: The Mystère-Falcon 50 provides essentially similar passenger accommodation to that of the twin-engined Mystère-Falcon 200 (see pages 80–81) but offers appreciably greater range. It features an entirely new wing of supercritical section and a structural fuel tank in the rear fuselage, aft of the pressure bulkhead, the latter having been adopted for the Mystère-Falcon 200 series, together with common systems and equipment.

DASSAULT-BREGUET MYSTÈRE-FALCON 50

Dimensions: Span, 61 ft 10½ in (18,86 m); length, 60 ft 9 in (18.52 m); height, 22 ft 10⅜ in (6,97 m); wing area, 504·13 sq ft (46,83 m²).

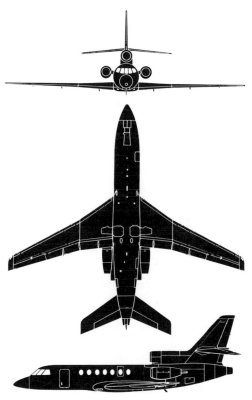

DASSAULT-BREGUET/DORNIER ALPHA JET

Countries of Origin: France and Federal Germany.
Type: Tandem two-seat basic/advanced trainer and light tactical aircraft.
Power Plant: Two 2,975 lb st (1 350 kgp) SNECMA-Turboméca Larzac 04-C5 turbofans.
Performance: Max. speed, 622 mph (1 000 km/h) or Mach 0·82 at sea level, 567 mph (912 km/h) or Mach 0·84 at 32,810 ft (10 000 m); max. climb. 11,220 ft/min (57 m/sec); ceiling, 45,000 ft (13 715 m); radius of action (training mission), 267 mls (430 km) at low altitude, 683 mls (1 100 km) at high altitude; ferry range (max. external fuel), 1,785 mls (2 872 km).
Weights: Empty, 7,716 lb (3 500 kg); loaded (clean), 11,023 lb (5 000 kg); max. take-off, 15.983 lb (7 250 kg).
Armament: (Close air support) External centreline gun pod with (Alpha Jet E) 30-mm DEFA 533 or (Alpha Jet A) 27-mm Mauser cannon, plus up to 4,850 lb (2 200 kg) of ordnance on four wing stations.
Status: First prototype flown 26 October 1973. Combined production rate of 10 monthly from French (Alpha Jet E) and German (Alpha Jet A) assembly lines, with some 390 delivered by beginning of 1982 against orders from France (175), Germany (175), Belgium (33), Cameroun (6), Egypt (30), Ivory Coast (6), Qatar (6), Morocco (24), Nigeria (16) and Togo (5). Final assembly lines in Toulouse and Munich.
Notes: French version is optimised for basic-advanced training and German version for close air support.

DASSAULT-BREGUET/DORNIER ALPHA JET

Dimensions: Span, 29 ft 11 in (9,11 m); length, 40 ft 3 in (12,29 m); height, 13 ft 9 in (4,19 m); wing area, 188 sq ft (17,50 m²).

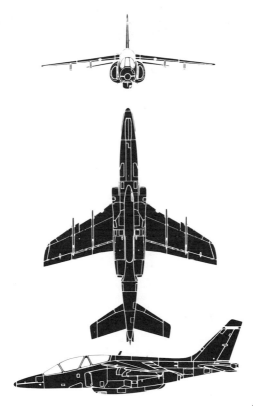

DE HAVILLAND CANADA DASH 7

Country of Origin. Canada.
Type: STOL short-haul commercial transport.
Power Plant: Four (flat-rated) 1,120 shp Pratt & Whitney (Canada) PT6A-50 turboprops.
Performance: Max. cruising speed (with 9,500-lb/4 309 kg payload of 50 passengers and baggage), 266 mph (428 km/h) at 8,000 ft (2 440 m); normal cruise, 262 mph (421 km/h) at 15,000 ft (4 570 m); range cruise, 248 mph (399 km/h); range (50 passengers and IFR reserves) at normal cruise, 795 mls (1 280 km), (max. fuel and 6,500 lb/2 948 kg payload), 1,347 mls (2 168 km).
Weights: Empty equipped, 27,600 lb (12 519 kg); max. take-off, 44,000 lb (19 958 kg).
Accommodation: Flight crew of two and standard seating for 50 passengers in four-abreast arrangement. Optional all-freight, mixed passenger/freight and convertible layouts available.
Status: First of two pre-series aircraft flown 27 March 1975, with first production aircraft following on 30 May 1977. Sixty delivered by beginning of 1982, when orders totalled 120 aircraft and production was running at three monthly (reducing to 2·5 monthly). Thirty-one scheduled to be delivered during 1982.
Notes: A maritime surveillance version, the Dash 7R Ranger, offers extended payload and range performance, two examples having been built for the Canadian Coast Guard. Several "stretched" versions of the Dash 7 were under consideration by the company at the beginning of 1982, including 60–64 passenger version with uprated PT6A engines and a 70–80 passenger version with PW120 or CT7 engines.

DE HAVILLAND CANADA DASH 7

Dimensions: Span, 93 ft 0 in (28,35 m); length, 80 ft 7¾ in (24,58 m); height, 26 ft 2 in (7,98 m); wing area, 860 sq ft (79,90 m²).

DORNIER DO 128-6

Country of Origin: Federal Germany.

Type: Light utility transport.

Power Plant: Two (flat-rated) 400 shp Pratt & Whitney (Canada) PT6A-110 turboprops.

Performance: Max. speed, 211 mph (340 km/h); long-range cruise, 159 mph (256 km/h); max. initial climb, 1,260 ft/min (6,4 m/sec); service ceiling, 28,150 ft (8 580 m); range (with max. fuel), 1,030 mls (1 658 km).

Weights: Operational empty, 5,600 lb (2 540 kg); max. take-off, 9,480 lb (4 300 kg).

Accommodation: Pilot and co-pilot/passenger on flight deck and nine passengers in individual seats with centre aisle in commuterliner arrangement, or 13 inward-facing folding seats.

Status: The prototype Do 128-6 was flown on 4 March 1980, with customer deliveries commencing 1981.

Notes: The Do 128 is a modernised version of the Do 28D-2 Skyservant, two versions of which are currently in production, the Do 128-2 with 380 shp Avco Lycoming IGSO-540-A1E piston engines and the turbine-engined Do 128-6 described and illustrated.

92

DORNIER DO 128-6

Dimensions: Span, 51 ft 1 in (15,55 m); length, 37 ft 5 in (11,41 m); height, 12 ft 9½ in (3,90 m); wing area, 312·2 sq ft (29,00 m²).

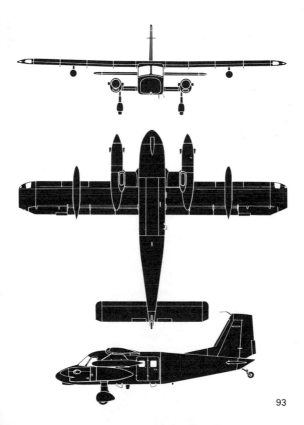

DORNIER DO 228-200

Country of Origin: Federal Germany.

Type: Light regional and utility transport.

Power Plant: Two (flat-rated) 715 shp Garrett AiResearch TPE 331-5 turboprops.

Performance: Max. cruising speed, 230 mph (370 km/h) at sea level, 268 mph (432 km/h) at 10,000 ft (3 280 m); long-range cruise, 206 mph (332 km/h) at 10,000 ft (3 280 m); range (19 passengers), 714 mls (1 150 km); initial climb, 2,050 ft/min (10,4 m/sec); service ceiling, 29,600 ft (9 020 m).

Weights: Operational empty, 7,370 lb (3 343 kg); max. take-off, 12,570 lb (5 700 kg).

Accommodation: Flight crew of two and standard arrangement for 19 passengers in individual seats with central aisle.

Status: Prototype Do 228-200 flown on 9 May 1981, preceded by prototype Do 228-100 on 28 March 1981. First of initial production batch of five Do 228-100s was scheduled to be delivered January 1982, and production rate of five aircraft monthly is planned to be attained in 1983.

Notes: The Do 228 mates a new-technology wing of supercritical section with the fuselage cross-section of the Do 128, and two versions are currently being manufactured, the Do 228-100, which has an overall length of 49 ft 3 in (15,03 m) and has four fewer seats, and the Do 228-200 illustrated and described. In other respects, the two versions are virtually identical and, apart from maximum range, the performances of the 15- and 19-passenger models are essentially similar. Both -100 and -200 have STOL capabilities, the new aerofoil offering considerably improved lift and the high-lift system combining single-slot Fowler flaps and "flaperons". The cabin has a constant rectangular cross-section.

DORNIER DO 228-200

Dimensions: Span, 55 ft 7 in (16,97 m); length, 54 ft 3 in (16,55 m); height, 15 ft 9 in (4,86 m); wing area, 344·46 sq ft (32,00 m²).

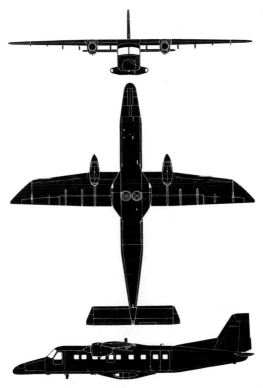

EMBRAER EMB-110/41 BANDEIRANTE

Country of Origin: Brazil.
Type: Short-haul commercial regional airliner.
Power Plant: Two (flat-rated) 750 shp Pratt & Whitney (Canada) PT6A-34 turboprops.
Performance: Max. cruising speed, 254 mph (409 km/h) at 10,000 ft (3 050 m); econ. cruise, 203 mph (326 km/h); max. initial climb, 1,660 ft/min (8,43 m/sec); ceiling, 21,400 ft (6 523 m); range (max. payload and 45 min reserves), 276 mls (445 km), (max. fuel and same reserves), 1,150 mls (1 852 km).
Weights: Empty equipped (P1), 7,855 lb (3 563 kg), (P2), 7,749 lb (3 515 kg); max. take-off, 13,007 lb (5 900 kg).
Accommodation: Pilot and co-pilot on flight deck, and (P1) quick-change seating for 18 passengers, or (P2) seating for up to 21 passengers in seven three-abreast rows.
Status: Prototype Bandeirante flown on 26 October 1968, and first EMB-110P (146th Bandeirante) on 3 May 1977, with deliveries commencing in following year. From mid-1981, all Bandeirantes delivered to FAA's new SFAR-41 regulations as EMB-110/41. Some 400 Bandeirantes (all versions but excluding EMB-311) completed by beginning of 1982, when production was running at six monthly.
Notes: The Bandeirante (Pioneer) has been the subject of continuous development, current production versions (primarily the EMB-110P1 and P2, the former having an enlarged cargo door) offering increased payload and range as a result of a series of minor changes conforming with SFAR-41 regulations. Retrofit to the new standards is possible to upgrade earlier production Bandeirantes.

EMBRAER EMB-110/41 BANDEIRANTE

Dimensions: Span, 50 ft 3⅛ in (15,32 m); length, 49 ft 5¾ in (15,08 m); height, 16 ft 1⅜ in (4,92 m); wing area, 312 sq ft (29,00 m²).

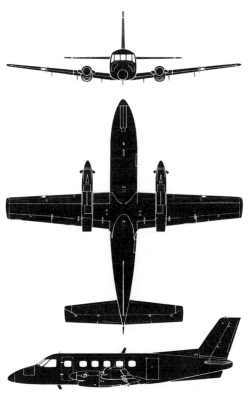

EMBRAER EMB–121A1 XINGU II

Country of Origin: Brazil.
Type: Light business executive transport and crew trainer.
Power Plant: Two 750 shp Pratt & Whitney (Canada) PT6A-135 turboprops.
Performance: Max. cruising speed (at 11,020 lb/5 000 kg), 298 mph (480 km/h) at 12,000 ft (3 660 m), 289 mph (465 km/h) at 20,000 ft (6 100 m); initial climb, 1,820 ft/min (9,24 m/sec); service ceiling, 28,000 ft (8 535 m); range (with max. payload and 45 min reserves), 1,059 mls (1 705 km), at long-range cruise, (max. fuel), 1,404 mls (2 260 km).
Weights: Basic empty, 8,200 lb (3 720 kg); max. take-off, 12,500 lb (5 670 kg).
Accommodation: Two seats side by side on flight deck and provision for up to nine passengers in main cabin.
Status: Prototype Xingu flown 10 October 1976, with first production aircraft following on 20 May 1977, and some 50 delivered by beginning of 1982, when production was being increased from 1·5 to 2·5 monthly to meet orders for 25 from France's Armée de l'Air and 16 for the Aéronavale for crew training. The more powerful Xingu II first flown on 4 September 1981 will replace initial model during course of 1982.
Notes: A developed version of the basic aircraft, the Xingu III with 850 shp PT6A-42 engines and a 25-in (63,5 cm) fuselage stretch, was under flight test at the beginning of 1982. The Xingu III has a longer wing span and discards the ventral fin.

EMBRAER EMB–121A1 XINGU II

Dimensions: Span, 47 ft 5 in (14,45 m); length, 40 ft 2¼ in (12,25 m); height, 15 ft 6½ in (4,74 m); wing area, 296 sq ft (27,50 m²).

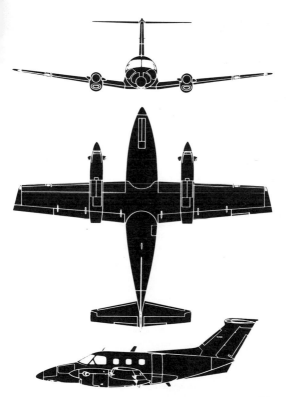

EMBRAER EMB-312 (T-27) TUCANO

Country of Origin: Brazil.
Type: Tandem two-seat basic and armament trainer.
Power Plant: One 750 shp Pratt & Whitney (Canada) PT6A-25C turboprop.
Performance: Max speed, 296 mph (476 km/h) at 7,000 ft (2135 m); max. continuous cruise, 270 mph (434 km/h); initial climb, 2,750 ft/min (13,97 m/sec); time to 25,000 ft (7620 m), 17 min; tactical radius (HI-LO-HI with two 250-lb/113-kg bombs), 403 mls (650 km), (with four 250-lb/113-kg bombs), 168 mls (270 km).
Weights: Empty, 3,748 lb (1700 kg); max. take-off, 5,180 lb (2350 kg).
Armament: Four underwing hardpoints for total weapons load of 1,323 lb (600 kg) which may comprise two 0·5-in (12,7-mm) gun pods, four pods each with seven 37-mm or 70-mm rockets, or four 250-lb (113-kg) bombs.
Status: First of two prototypes of the Tucano was flown on 16 August 1980, and 112 ordered for the Brazilian Air Force with a further 50 on option. Initial production deliveries scheduled for summer 1982, with peak production tempo of five aircraft monthly scheduled to be attained during 1983.
Notes: The T-27 Tucano, which is unique among current turboprop-powered basic trainers in having ejection seats, has been designed to simulate the flying characteristics of a pure jet aircraft, thus enabling the Brazilian Air Force to omit a jet trainer from its basic wings course. When the Tucano enters service in 1983, pupil pilots will fly 120 hours on this type following 50 hours on a primary trainer such as the T-17 Tangará and 50 hours in a Tucano simulator. The Tucano is also suitable for armament training and for the light counter-insurgency role.

EMBRAER EMB-312 (T-27) TUCANO

Dimensions: Span, 36 ft 6½ in (11,14 m); length, 32 ft 4¼ in (9,86 m); height, 11 ft 1⅞ in (3,40 m); wing area, 204·52 sq ft (19,00 m²).

FAIRCHILD A-10A THUNDERBOLT II

Country of Origin: USA.

Type: Single-seat close air support aircraft.

Power Plant: Two 9,065 lb st (4 112 kgp) General Electric TF34-GE-100 turbofans.

Performance: Max. speed, 433 mph (697 km/h) at sea level, 448 mph (721 km/h) at 10,000 ft (3 050 m); initial climb, 5,340 ft/min (27,12 m/sec); service ceiling, 34,700 ft (10 575 m); combat radius (with 9,540 lb/4 327 kg bomb load and loiter allowance), 288 mls (463 km) at (average) 329 mph (529 km/h) at 25,000–35,000 ft (7 620–10 670 m); ferry range, 2,487 mls (4 000 km).

Weights: Empty, 19,856 lb (9 006 kg); basic operational, 22,844 lb (10 362 kg); max. take-off, 46,786 lb (22 221 kg).

Armament: One seven-barrel 30-mm GAU-8 rotary cannon and max. of 16,000 lb (7 250 kg) of external ordnance on 11 stations, or, with max. internal fuel and cannon ammunition, 9,540 lb (4 327 kg).

Status: First of two prototypes flown 10 May 1972, with first production A-10A following six pre-production aircraft on 21 October 1975. Current planning calls for total of 825 aircraft with completion of deliveries in April 1986, and including (from Fiscal 1981 funding) 30 two-seat A-10B combat readiness trainers, with deliveries commencing 1982.

Notes: The A-10A dedicated close air support aircraft is being delivered to the USAF, Air National Guard and Air Force Reserve. The A-10B (the two-seat airframe of which is based on that of the experimental night and adverse weather version described in the 1980 edition) has 94% structural commonality with the A-10A, and will be capable of combat missions (with same fuel and stores as the A-10A) when flown as a single-seater.

FAIRCHILD A-10A THUNDERBOLT II

Dimensions: Span, 57 ft 6 in (17,53 m); length, 53 ft 4 in (16,25 m); height, 14 ft 8 in (4,47 m); wing area, 506 sq ft (47,01 m²).

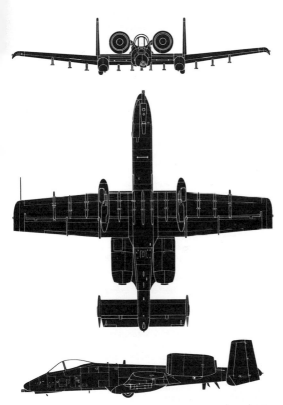

FAIRCHILD SWEARINGEN METRO IIIA

Country of Origin: USA.

Type: Short-haul regional transport.

Power Plant: Two (Metro III) 1,000 shp Garrett AiResearch TPE 331-11U-601G or (Metro IIIA) 956 shp Pratt & Whitney (Canada) PT6A-45R turboprops.

Performance: (Metro III) Max. speed, 345 mph (523 km/h); max. cruise, 319 mph (513 km/h) at 10,000 ft (3 050 m); initial climb, 2,440 ft/min (12,4 m/sec); service ceiling, 30,000 ft (9 150 m); range (19 passengers and 45 min reserves), 714 mls (1 149 km).

Weights: (Metro III) Operational empty, 8,737 lb (3 963 kg); max. take-off, 14,000 lb (6 350 kg); (Metro IIIA) max. take-off, 14,500 lb (6 577 kg).

Accommodation: Flight crew of two and standard arrangement for 19 passengers seated two abreast.

Status: The Metro III was first flown in 1980, with first customer deliveries mid-1981, with 20 by year's end. The Metro IIIA was scheduled to commence flight test late 1981, with initial customer deliveries planned for late 1983.

Notes: The Metro III and IIIA (and the corporate executive transport equivalent of the former, known as the Merlin IVC) differ structurally from previous Metros in having a new, longer span wing which is mated with uprated engines to cater for increased gross weights. The Metro was first flown on 26 August 1969, entering service in 1971, a series of design improvements resulting in the Metro II in 1975. All Metros prior to the Metro IIIA (illustrated on opposite page) have been powered by Garrett AiResearch engines, Pratt & Whitney engines being adopted for the new model to provide a customer option. More than 200 Metros have been sold to over 40 operators, and production was running at four–five monthly at the beginning of 1982.

FAIRCHILD SWEARINGEN METRO IIIA

Dimensions: Span, 57 ft 0 in (17,37 m); length, 59 ft 4¼ in (18,09 m); height, 16 ft 8 in (5,08 m); wing area, 309 sq ft (28,71 m²).

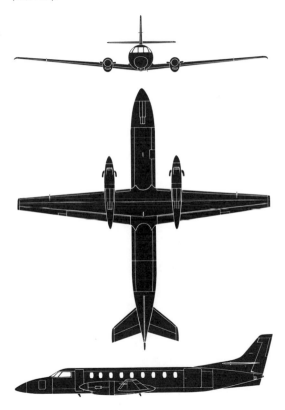

FOKKER F27 FRIENDSHIP Mк 500

Country of Origin: Netherlands.

Type: Short/medium-haul commercial transport.

Power Plant: Two 2,140 shp Rolls-Royce Dart Mk 536-7R turboprops.

Performance: Max. cruising speed, 298 mph (480 km/h) at 20,000 ft (6 095 m); initial climb, 1,480 ft/min (7,52 m/sec); service ceiling, 29,500 ft (8 990 m); range (max. payload and reserves of 10% plus 30 min hold), 1,082 mls (1 741 km).

Weights: Operational empty, 26,345 lb (11 950 kg); max. take-off, 45,000 lb (20 412 kg).

Accommodation: Flight crew of two or three on flight deck and provision for 52 passengers four abreast in main cabin with alternative arrangements for up to 60 passengers.

Status: First of two F27 prototypes flown on 24 November 1955, with first production deliveries following in November 1958. Total of 739 F27s ordered by the beginning of 1982 (including 205 built in USA by Fairchild).

Notes: Current production versions of the F27 include the F27MP maritime surveillance aircraft (see 1981 edition) which can be based on either the Mk 200 or Mk 400 airframe, the latter having a forward freight door and reinforced floor, the Mk 500, which is essentially similar to the Mk 200 apart from a 4 ft 11 in (1,5 m) fuselage stretch, and the Mk 600 which is similar to the Mk 200 but has a large cargo door. The F27MP features surveillance radar, long-range inertial navigation equipment, blister observation windows and provision for wing pylon tanks. It has been purchased by the Netherlands, Angola, Peru, the Philippines and Spain.

FOKKER F27 FRIENDSHIP Mᴋ 500

Dimensions: Span, 95 ft 1¾ in (29,00 m); length, 82 ft 2½ in (25,06 m); height, 28 ft 7¼ in (8,71 m); wing area, 753·47 sq ft (70,00 m²).

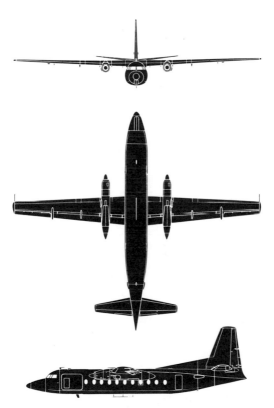

FOKKER F28 FELLOWSHIP Mк 4000

Country of Origin: Netherlands.

Type: Short/medium-haul commercial transport.

Power Plant: Two 9,850 lb st (4468 kgp) Rolls-Royce RB. 183-2 Spey Mk 555-15H turbofans.

Performance: Max. cruising speed, 523 mph (843 km/h) at 23,000 ft (7000 m); econ. cruise, 487 mph (783 km/h) at 32,000 ft (9755 m); range cruise, 421 mph (678 km/h) at 30,000 ft (9145 m); range (with max. payload), 1,160 mls (1870 km), (with max. fuel), 2,566 mls (4130 km); cruise altitude, 35,000 ft (10675 m).

Weights: Operational empty, 38,825 lb (17661 kg); max. take-off, 73,000 lb (33110 kg).

Accommodation: Flight crew of two on flight deck (with jump seat for third crew member) and basic main cabin single-class configuration for 85 passengers seated five abreast.

Status: First of two F28 prototypes flown on 8 May 1967, with first customer delivery following on 24 February 1969. Total of 189 F28s had been ordered by 46 operators in 31 countries by the beginning of 1982.

Notes: The F28 Mks 1000 and 2000 are now out of production (after completion of 97 and 10 respectively), having been replaced by the Mks 3000 and 4000, and these remained the current production versions at the beginning of 1982. The Mk 3000 has the 80 ft 6½ in (24,55 m) fuselage of the Mk 1000 offering seating for up to 65 passengers, and the Mk 4000 has the lengthened fuselage introduced by the Mk 2000. Both versions have the extended wing originally developed for the slat-equipped Mks 5000 and 6000 (now discarded) but with fixed leading edges.

FOKKER F28 FELLOWSHIP Mᴋ 4000

Dimensions: Span, 82 ft 3 in (25,07 m); length, 97 ft 1¾ in (29,61 m); height, 27 ft 9½ in (8,47 m); wing area, 850 sq ft (78,97 m²).

GAF NOMAD N22B

Country of Origin: Australia.

Type: Light STOL utility transport.

Power Plant: Two 400shp Allison 250-B17B turboprops.

Performance: Max. cruising speed, 193 mph (311 km/h) at 5,000 ft (1 525 m); normal cruise (90% power), 190 mph (306 km/h), (75% power), 176 mph (283 km/h); initial climb, 1,460 ft/min (7,4 m/sec); service ceiling, 23,500 ft (7 165 m); range (with 45 min reserves at econ. cruise), 920 mls (1 480 km).

Weights: Operational empty, 4,783 lb (2 170 kg); max. take-off, 8,500 lb (3 856 kg).

Accommodation: Flight crew of one or two on flight deck with provision for up to 13 passengers in individual seats in main cabin, or aeromedical arrangement for two casualty stretchers, three seated casualties and a medical attendant.

Status: The prototype Nomad was flown on 23 July 1971, with first production aircraft flying in May 1975, and first commercial delivery (N22B) following in September 1975. First stretched version (N24A) flown July 1975, with approximately 120 (all versions) delivered by beginning of 1982 in a ratio of roughly one N24A to three N22Bs. Production batches totalling 145 aircraft committed with follow-on batch of a further 55 planned at beginning of 1982.

Notes: The Nomad is being manufactured for a variety of roles, ranging from 12- and 16-passenger regional airliner (Nomad N22B and N24A Commuterliner) and military support aircraft (N22B Missionmaster) to Searchmaster "B" and "L" maritime surveillance versions of the N22B. A twin-float version of the N22B is known as the Nomad Floatmaster.

110

GAF NOMAD N22B

Dimensions: Span, 54 ft 0 in (16,46 m); length, 41 ft 2½ in (12,57 m); height, 18 ft 1½ in (5,52 m); wing area, 324 sq ft (30,10 m²).

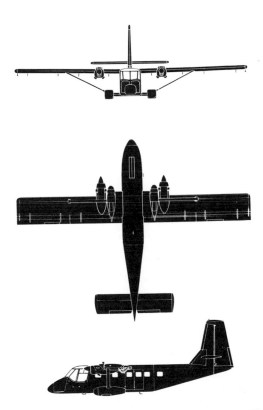

GATES LEARJET 35A

Country of Origin: USA.

Type: Light business executive transport.

Power Plant: Two 3,500 lb st (1 588 kgp) Garrett AiResearch TFE 731-2-2B turbofans.

Performance: Max. speed, 542 mph (872 km/h) at 25,000 ft (7 620 m); max. cruise, 528 mph (850 km/h) at 41,000 ft (12 500 m); range (four passengers and 45 min reserves), 2,789 mls (4 488 km); max. initial climb, 4,900 ft/min (24,89 m/sec); service ceiling, 45,000 ft (13 715 m).

Weights: Empty equipped, 9,571 lb (4 341 kg); max. take-off, 17,000 lb (7 711 kg).

Accommodation: Crew of two on flight deck and maximum of eight passengers in main cabin.

Status: The Learjet 35A appeared in 1976 as an upgraded version of the original Learjet 35, which, itself a turbofan-powered growth version of the Learjet 25, flew as a prototype on 22 August 1973, customer deliveries commencing November 1974. Production of the Learjet 35A (and the smaller-capacity longer-range 36A) was approximately seven per month at the beginning of 1982.

Notes: The Learjet 35A and 36A are essentially similar excepting that the latter has a larger fuselage tank which reduces main cabin accommodation to a maximum of six passengers, maximum range with four passengers and 45 min reserves being increased by 500 mls (805 km). A special-missions version of the 35A (mapping, target-towing, sea patrol, etc) has been ordered by Finland (three) for 1982 delivery.

GATES LEARJET 35A

Dimensions: Span, 39 ft 6 in (12,04 m); length, 48 ft 8 in (14,33 m); height, 12 ft 3 in (3,73 m); wing area, 253·3 sq ft (23,50 m²).

GATES LEARJET LONGHORN 55

Country of Origin: USA.

Type: Light business executive transport.

Power Plant: Two 3,700 lb st (1 678 kgp) Garrett AiResearch TFE 731-3A-2B turbofans.

Performance: Max. cruising speed, 525 mph (845 km/h) at 41,000 ft (12 495 m); max. range (with four passengers and 45 min reserves), 2,492 mls (4 010 km) at 475 mph (765 km/h) at 41,000 ft (12 495 m); time to 41,000 ft (12 495 m), 24 min.

Weights: Empty equipped, 12,130 lb (5 501 kg); max. take-off, 19,500 lb (8 845 kg), (optional), 20,500 lb (9 299 kg).

Accommodation: Flight crew of two and various main cabin arrangements for up to a maximum of 11 passengers.

Status: The first of two Longhorn 55 prototypes flown on 19 April 1979, with first production aircraft following on 11 August 1980. Customer deliveries commenced August 1981, and 19 were scheduled to be delivered by the beginning of 1982 when approximately 165 were on order. Planned production rate of five aircraft monthly in second quarter of 1982.

Notes: The Longhorn 55 mates the wing development of the Longhorn 28/29 (see 1979 edition) with an entirely new fuselage possessing some 50% greater cross-section. With the optional higher max. take-off weight, max. range with four passengers is increased to 2,666 mls (4 290 km). A longer-range Longhorn 56 with a smaller cabin and maximum capacity for eight passengers is proposed but has not yet been scheduled for FAA certification. Consideration is being given to the application of the 4,300 lb st (1 950 kgp) TFE 731-5 engine for installation from 1984.

114

GATES LEARJET LONGHORN 55

Dimensions: Span, 43 ft 9½ in (13,34 m); length, 55 ft 1½ in (16,79 m); height, 14 ft 8 in (4,47 m); wing area, 264·5 sq ft (24,57 m²).

GENERAL DYNAMICS F-16
FIGHTING FALCON

Country of Origin: USA.

Type: Single-seat multi-role fighter (F-16A) and two-seat operational trainer (F-16B).

Power Plant: One 14,800 lb st (6 713 kgp) dry and 23,830 lb st (10 809 kgp) reheat Pratt & Whitney F100-PW-200 turbofan.

Performance: Max speed (short endurance dash), 1,333 mph (2 145 km/h) or Mach 2·02, (sustained), 1,247 mph (2 007 km/h) or Mach 1·89 at 40,000 ft (12 190 m); max. cruise, 614 mph (988 km/h) or Mach 0·93; tactical radius (HI-LO-HI interdiction on internal fuel), 360 mls (580 km) with six 500-lb (227-kg) Mk 82 bombs; range (internal fuel and similar ordnance load), 1,200 mls (1 930 km) at (average) 575 mph (925 km/h); ferry range (max. external fuel), 2,535 mls (4 080 km).

Weights: Operational empty, 14,567 lb (6 613 kg); max. take-off, 35,400 lb (16 057 kg).

Armament: One 20-mm M61A-1 multi-barrel rotary cannon and from two to six AIM-9L/M Sidewinder AAMs, or (air support) up to 12,000 lb (5 443 kg) of ordnance distributed between nine stations.

Status: First of two (YF-16) prototypes flown 20 January 1974, and first of eight FSD (Full Scale Development) aircraft following 8 December 1976. First production F-16 flown 7 August 1978, with some 560 delivered by beginning of 1982 by parent company (10 monthly) and European consortium (five monthly) with final assembly lines in Netherlands and Belgium. Enlarged tailplane (see drawing) introduced from November 1981. Single-seat F-16C and two-seat F-16D with upgraded systems to be delivered from July 1984.

116

GENERAL DYNAMICS F-16 FIGHTING FALCON

Dimensions: Span (excluding missiles), 31 ft 0 in (9,45 m); length, 47 ft 7¾ in (14,52 m); height, 16 ft 5¼ in (5,01 m); wing area, 300 sq ft (27,87 m²).

GENERAL DYNAMICS F-16E

Country of Origin: USA.

Type: Single- (or two-) seat advanced multi-role fighter technology demonstrator.

Power Plant: One 14,800 lb st (6 713 kgp) dry and 23,830 lb st (10 809 kgp) reheat Pratt & Whitney F100-PW-200 turbofan.

Performance: (Estimated) Max. speed (short endurance dash), 1,650 mph (2 655 km/h) or Mach 2·5 at 40,000 ft (12 190 m); max. cruise, 1,452 mph (2 337 km/h) or Mach 2·2 above 40,000 ft (12 190 m); tactical radius (HI-LO-HI interdiction mission on internal fuel with 8,000 lb/3 629 kg ordnance and allowance for combat), 520 mls (837 km), (with 4,000 lb/1 814 kg ordnance), 805 mls (1 295 km).

Weights: (Estimated) Empty, 17,402 lb (7 894 kg); max. takeoff, 37,500 lb (17 010 kg).

Armament: (Air combat) Up to 10 AMRAAM (Advanced Medium Range Air-to-Air Missile), or (air support) 22 500-lb (227-kg) Mk 82 bombs, or six AGM-65 Maverick missiles.

Status: Two FSD (Full Scale Development) F-16 airframes were being rebuilt to F-16E configuration (one single- and one two-seat) at the beginning of 1982, with the first being scheduled to commence flight test in July 1982.

Notes: Featuring a wing of so-called "cranked arrow" configuration, the F-16E has 72% airframe and sub-systems commonality with the basic Fighter Falcon (see pages 116–17) but will offer major improvements in field performance, tactical radius, manœuvre envelope, penetration speed and payload.

118

GENERAL DYNAMICS F-16E

Dimensions: Span, 32 ft 5 in (9,87 m); length, 52 ft 5 in (15,97 m); height, 16 ft 5 in (5,00 m); wing area, 646·4 sq ft (60,05 m²).

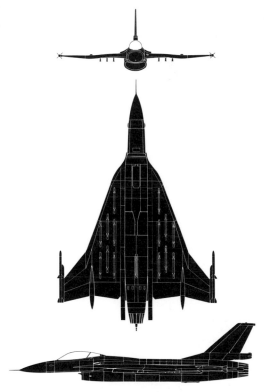

GRUMMAN E-2C HAWKEYE

Country of Origin: USA.
Type: Airborne early warning, surface surveillance and strike control aircraft.
Power Plant: Two 4,910 ehp Allison T56-A-425 turboprops.
Performance: Max. speed, 348 mph (560 km/h) at 10,000 ft (3 050 m); max. range cruise, 309 mph (498 km/h); initial climb, 2,515 ft/min (12,8 m/sec); service ceiling, 30,800 ft (9 390 m); mission endurance (at 230 mls/370 km from base), 4·0 hrs; max. endurance, 6·1 hrs; ferry range, 1,604 mls (2 580 km).
Weights: Empty, 38,009 lb (17 240 kg); max. take-off, 51,900 lb (23 540 kg).
Accommodation: Crew of five comprising flight crew of two and Airborne Tactical Data System team of three, each occupying an independent operating station.
Status: First of two E-2C prototypes flown on 20 January 1971, with first production aircraft flying on 23 September 1972. Total US Navy requirement for 101 by 1986, with some 70 delivered by the beginning of 1982, in addition to four to Israel. Eight E-2Cs ordered by Japan with deliveries commencing 1982, and the procurement of four being negotiated by Egypt for 1983 delivery
Notes: The E-2C followed 59 E-2As (all subsequently updated to E-2B standards) and is able to operate independently, in co-operation with other aircraft, or in concert with ground environments. Two examples of a training version, the TE-2C, have been delivered to the US Navy, and late 1982, the procurement of the E-2C was being considered by France. Production of the E-2C is scheduled to continue through 1986.

GRUMMAN E-2C HAWKEYE

Dimensions: Span, 80 ft 7 in (24,56 m); length, 57 ft 7 in (17,55 m); height, 18 ft 4 in (5,69 m); wing area, 700 sq ft (65,03 m²).

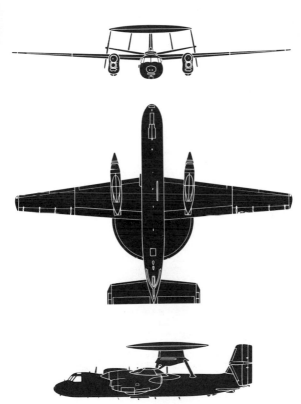

GRUMMAN F-14A TOMCAT

Country of Origin: USA.
Type: Two-seat shipboard multi-role fighter.
Power Plant: Two 12,500 lb st (5 670 kgp) dry and 20,900 lb st (9 480 kgp) reheat Pratt & Whitney TF30-P-412A turbofans.
Performance: Max. speed (with four semi-recessed AIM-7 AAMs), 913 mph (1 470 km/h) or Mach 1·2 at sea level, 1,584 mph (2 549 km/h) or Mach 2·4 at 49,000 ft (14 935 m); time to 60,000 ft (18 290 m) at 55,000 lb (24 948 kg), 2·1 min; tactical radius (combat air patrol on internal fuel with full complement of AAMs), 765 mls (1 232 km), (interdiction mission with HI-LO-HI profile with drop tanks and 7,000 lb/3 175 kg of bombs), 725 mls (1 167 km).
Weights: Empty, 39,930 lb (18 112 kg); loaded (intercept mission with four AIM-7s), 58,904 lb (26 718 kg), (with six AIM-54s), 69,790 lb (31 656 kg); max., 74,348 lb (33 724 kg).
Armament: One 20-mm M61A-1 rotary cannon and (intercept) six AIM-7E/F Sparrow and four AIM-9G/H Sidewinder AAMs, or six AIM-54A Phoenix and two AIM-9G/H AAMs.
Status: First of 12 R&D aircraft flown 21 December 1970, and some 350 F-14As in US Navy service at the beginning of 1982, when development included the enhanced F-14C for 1983 delivery and the F-14D which will be introduced (according to US Navy advanced planning) after delivery of 605 Tomcats against proposed total of 845.
Notes: New Tomcats for 1983 delivery will be of the F-14C version with the enhanced TF30-PW-414A engine and other improvements (existing F-14As being modified to similar standards from 1984), and the F-14D will feature the General Electric F101 engine and a possible fuselage "stretch".

GRUMMAN F-14A TOMCAT

Dimensions: Span (20 deg sweep), 64 ft 1½ in (19,55 m); (68 deg sweep) 37 ft 7 in (11,45 m); length, 61 ft 11⅞ in (18,90 m); height, 16 ft 0 in (4,88 m); wing area, 565 sq ft (52,50 m²).

GULFSTREAM AMERICAN
GULFSTREAM III

Country of Origin: USA.

Type: Light business executive transport.

Power Plant: Two 11,400 lb st (5171 kgp) Rolls-Royce RB.163-25 Spey Mk 511-8 turbofans.

Performance: Max. cruising speed, 577 mph (928 km/h) or Mach 0·85; long-range cruise, 512 (825 km/h) or Mach 0·775; max. operational altitude, 45,000 ft (13 715 m); max. range, 5,129 mls (8 255 km), (with four passengers and IFR reserves at Mach 0·8), 4,254 mls (6 846 km), (with 18 passengers), 3,930 mls (6 327 km).

Weights: Operational empty (typical), 38,300 lb (17 374 kg); max. take-off, 68,200 lb (30 936 kg).

Accommodation: Flight crew of two and various main cabin arrangements for 8–12 passengers in executive transport configuration. Maximum high-density configuration for 19 passengers.

Status: Gulfstream III prototype first flown 2 December 1979, with customer deliveries commencing September 1980, and 40 delivered by beginning of 1982 against total orders for more than 100 and a production rate of three monthly.

Notes: The Gulfstream III is a progressive development of the (Grumman) Gulfstream II (see 1969 edition) of which 258 were built, including the two airframes utilised as Gulfstream III prototypes. Delivery of three to the Danish Air Force as special-role aircraft commenced late 1981, these having a large cargo door in the forward fuselage.

124

GULFSTREAM AMERICAN GULFSTREAM III

Dimensions: Span, 77 ft 10 in (23,72 m); length, 83 ft 1 in (25,30 m); height, 24 ft 4½ in (7,40 m); wing area, 934·6 sq ft (86,82 m²).

GULFSTREAM AMERICAN
PEREGRINE

Country of Origin: USA.

Type: Side-by-side two-seat basic trainer.

Power Plant: One 3,000 lb st (1 360 kgp) Pratt & Whitney (Canada) JT15D-5 turbofan.

Performance: (Estimated for proposed production model) Max. speed, 441 mph (710 km/h) at 30,000 ft (9 145 m); average cruise, 425 mph (684 km/h) at 30,000 ft (9 145 m); time to 40,000 ft (12 190 m), 16·5 min; service ceiling, 48,000 ft (14 630 m); range (internal fuel), 1,243 mls (2 000 km).

Weights: Max. loaded, 6,200 lb (2 812 kg).

Status: Prototype flown on 22 May 1981. Production commitment expected first quarter of 1983, with a view to deliveries commencing 1984.

Notes: Based broadly on the design of the push-pull Hustler twin-engined business aircraft (see 1980 edition), development of which was dormant at the beginning of 1982, the Peregrine is being offered with a pair of 1,500 lb st (680 kgp) Williams International WR-44 turbofans as an alternative installation to the single JT15D-5, and with tandem in place of side-by-side seating. The prototype (illustrated) is powered by a 2,500 lb st (1 134 kgp) JT15D-4 turbofan, with which a max. speed of 393 mph (632 km/h) is anticipated, and the Peregrine is claimed to offer a fuel consumption some 30% better than achieved by current jet trainers of comparable size.

126

GULFSTREAM AMERICAN PEREGRINE

Dimensions: Span, 34 ft 5½ in (10,50 m); length, 38 ft 4 in (11,68 m); height, 13 ft 5 in (4,09 m).

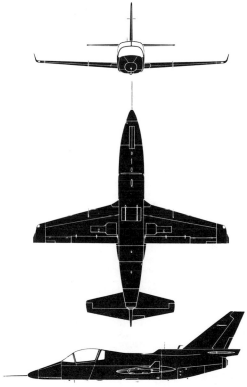

IAI KFIR-TC2

Country of Origin: Israel.

Type: Tandem two-seat conversion trainer.

Power Plant: One 11,870 lb st (5 385 kgp) dry and 17,900 lb st (8 120 kgp) reheat Bet-Shemesh-built General Electric J79-GE-17 turbojet.

Performance: (Estimated) Max. speed, 850 mph (1 368 km/h) or Mach 1·12 at 1,000 ft (305 m), 1,420 mph (2 285 km/h) or Mach 2·3 above 36,000 ft (10 970 m); max. climb, 47,250 ft/min (240 m/sec); ceiling, 59,050 ft (18 000 m).

Weights: (Estimated) Normal loaded, 20,944 lb (9 500 kg).

Armament: Two 30-mm DEFA cannon with 125 rpg. Hardpoints for up to 8,820 lb (4 000 kg) of external ordnance for secondary close air support mission.

Status: First Kfir-TC2 flown in February 1981, and production for Israeli Air Force continuing at beginning of 1982 in parallel with single-seat Kfir-C2.

Notes: A two-seat derivative of the Kfir-C2 multi-role fighter (see 1981 edition), the Kfir-TC2 has reduced internal fuel capacity resulting from insertion of a second cockpit between the air intake trunks, a 2 ft 9 in (0,84 m) plug inserted in the forward fuselage, recontoured upper fuselage decking and a drooped nose. Pupil and instructor are accommodated in vertically staggered ejection seats and both cockpits are enclosed by a single aft-hinging canopy. The Kfir-TC2 is produced on the same assembly line as the single-seat model.

128

IAI KFIR-TC2

Dimensions: Span, 26 ft 11½ in (8,22 m); length, 53 ft 9¼ in (16,39 m); height, 13 ft 11½ in (4,25 m); wing area (excluding canard and dog-tooth), 375·12 sq ft (34,85 m²).

129

IAI WESTWIND II

Country of Origin: Israel.

Type: Light business executive transport.

Power Plant: Two 3,700 lb st (1 678 kgp) Garrett AiResearch TFE 731-1G turbofans.

Performance: Max. speed, 533 mph (858 km/h) at 29,000 ft (8 840 m); econ. cruise, 449 mph (723 km/h) at 39,000–41,000 ft (11 890–12 500 m); range (four passengers), 3,345 mls (5 383 km), (10 passengers), 2,752 mls (4 429 km); max. initial climb, 5,000 ft/min (25,39 m/sec).

Weights: Empty equipped, 13,250 lb (6 010 kg); max. take-off, 23,500 lb (10 660 kg).

Accommodation: Two seats side-by-side on flight deck with various arrangements in main cabin for 7–10 passengers.

Status: The Westwind II was first flown on 24 April 1979, sales exceeding 50 by the beginning of 1982, when production combined with that of the Westwind I at a rate of five monthly.

Notes: The Westwind II is a longer-range derivative of the Westwind I with a higher-efficiency aerofoil, winglets attached to the wingtip tanks and various other refinements. The Westwind I, which succeeded the Westwind 1124 (the first of the Westwind series with TFE 731 turbofans), offered increases in fuel and cabin capacity, and some 180 TFE 731-powered Westwinds had been delivered by the beginning of 1982, when metal was being cut on a further development of the basic design. This, a Mach 0·8 long-range aircraft known as the Astra, is intended for 1984 delivery and will mate the fuselage, power plant and tail unit of the Westwind II with an entirely new wing mounted in low position.

IAI WESTWIND II

Dimensions: Span, 44 ft 9½ in (13,65 m); length, 52 ft 3 in (15,93 m); height, 15 ft 9½ in (4,81 m); wing area, 308·26 sq ft (28,64 m²).

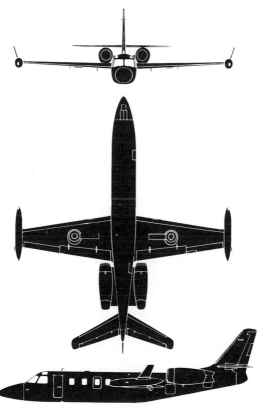

IAR-827A

Country of Origin: Romania.

Type: Side-by-side two-seat heavy-duty agricultural aircraft.

Power Plant: One 600 hp PZL-3S seven-cylinder radial air-cooled engine.

Performance: Max. speed, 130 mph (210 km/h); typical operating speed, 84–112 mph (135–180 km/h); initial climb rate, 590 ft/min (3,0 m/sec); normal operating range, 217 mls (350 km); service ceiling, 11,475 ft (3 500 m); max. endurance, 2·5 hrs.

Weights: Standard empty, 3,638 lb (1 650 kg); max. take-off, 6,173 lb (2 800 kg).

Status: The IAR-827A was first flown in 1979 (as a derivative of the basic IAR-827 flown in prototype form in July 1976), and the first production example was completed in the spring of 1980, production averaging one monthly in 1981.

Notes: The IAR-827A heavy-duty agricultural monoplane has been developed from the original IAR-827 prototype powered by a 400 hp Avco Lycoming IO-720-DA1B horizontally-opposed engine. Among the largest of the current generation of agricultural aircraft, the IAR-827A is unusual in providing side-by-side seating for two in the cockpit. Dual control can be provided to facilitate training, but the starboard seat is normally occupied by a mechanic/ground handler during ferry flights to operating zones. The hopper of the IAR-827A can carry up to 2,200 lb (1 000 kg), but normal operating loads are 1,764 lb (800 kg) dry or 264 Imp gal (1 200 l) of liquids. The airframe is claimed to have a life of 4,000 hours which is equivalent to 22,000 flights.

IAR-827A

Dimensions: Span, 45 ft 11 in (14,00 m); length, 28 ft 2½ in (8,60 m); height, 12 ft 3½ in (3,75 m); wing area, 322·9 sq ft (30,00 m²).

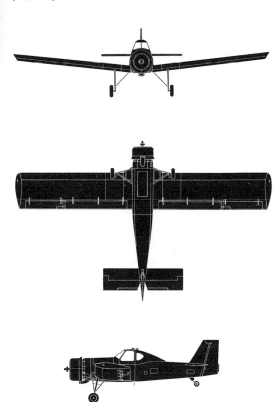

ILYUSHIN IL-76 (CANDID)

Country of Origin: USSR.

Type: Heavy-duty medium/long-haul freighter and troop transport.

Power Plant: Four 26,455 lb st (12 000 kgp) Soloviev D-30KP turbofans.

Performance: Max. speed, 528 mph (850 km/h) at 32,810 ft (10 000 m); max. cruise, 497 mph (800 km/h) at 29,500–42,650 ft (9 000–13 000 m); range cruise, 466 mph (750 km/h); range (with max. payload), 3,290 mls (5 300 km), (with max. fuel), 4,163 mls (6 700 km).

Weights: Max. take-off, 374,790 lb (170 000 kg).

Accommodation: Normal flight crew of four (with navigator below flight deck in glazed nose). Pressurised hold for up to 140 fully-equipped troops, containerised freight (up to 88,185 lb/ 40 000 kg), wheeled or tracked vehicles, self-propelled anti-aircraft guns (e.g., one radar-controlled ZSU-57-2 twin gun with seven-man crew or two ZSU-23-4 quad-guns with four-man crews), or mobile surface-to-air missile systems (e.g., SA-8 Gecko- or SA-9 Gaskin-equipped vehicles).

Armament: Twin 23-mm cannon in tail barbette.

Status: First of four prototypes flown on 25 March 1971, with production deliveries to both Soviet Air Force and Aeroflot commencing in 1974. Soviet Air Force air transport component had received 130-140 Il-76s by beginning of 1982, when production was running at 30–35 annually. Small numbers have been supplied to Iraq, Libya and Syria for dual military/civil tasks. In 1981, the Il-76 was selected to replace the An-12 by the Indian Air Force.

ILYUSHIN IL-76 (CANDID)

Dimensions: Span, 165 ft 8¼ in (50,50 m); length, 152 ft 10¼ in (46,59 m); height, 48 ft 5⅛ in (14,76 m); wing area, 3,229·2 sq ft (300,00 m²).

ILYUSHIN IL-86 (CAMBER)

Country of Origin: USSR.

Type: Medium-haul commercial transport.

Power Plant: Four 28.660 lb st (13 000 kqp) Kuznetsov NK-86 turbofans.

Performance: Max. cruising speed, 590 mph (950 km/h) at 29,530 ft (9 000 m); econ. cruise, 559 mph (900 km/h) at 36,090 ft (11 000 m); range (with max. payload—350 passengers), 2,485 mls (4 000 km), (with 250 passengers), 3,107 mls (5 000 km).

Weights: Max. take-off, 454,150 lb (206 000 kg).

Accommodation: Basic flight crew of three–four and up to 350 passengers in nine-abreast seating (with two aisles) divided between three cabins accommodating 111, 141 and 98 passengers respectively.

Status: First prototype flown on 22 December 1976, and production prototype flown on 24 October 1977. First scheduled service (Moscow–Tashkent) flown by Aeroflot on 26 December 1980, and first international service (Moscow–Prague) on 12 October 1981. Polish WSK-Mielec concern is responsible for the entire wing, stabiliser and engine pylons. Four are scheduled to be delivered to Polish Airlines LOT during 1982.

Notes: During two flights, on 22 and 24 September 1981, the Il-86 is claimed to have raised 18 international records by carrying a 144,594 lb (65 588 kg) load over a 1,242·7-mile (2 000-km) closed route at an average speed of 594 mph (956 km/h), and a 176,367 lb (80 000 kg) load over a 621·4-mile (1 000-km) route at 603 mph (971 km/h).

ILYUSHIN IL-86 (CAMBER)

Dimensions: Span, 157 ft 8⅛ in (48,06 m); length, 195 ft 4 in (59,54 m); height, 51 ft 10½ in (15,81 m); wing area, 3,550 sq ft (329,80 m²).

LEARAVIA LEAR FAN 2100

Country of Origin: USA.

Type: Light business executive aircraft.

Power Plant: Two 850 shp (flat-rated at 650 shp) Pratt & Whitney (Canada) PT6B-35F turboshafts.

Performance: Max. speed, 425 mph (684 km/h) at 20,000 ft (6 095 m); max. cruise, 418 mph (673 km/h) at 20,000 ft (6 093 m), 358 mph (576 km/h) at 40,000 ft (12 190 m); range cruise, 267 mph (430 km/h) at 20,000 ft (6 095 m), 322 mph (518 km/h) at 40,000 ft (12 190 m); initial climb (at max. take-off), 3,450 ft/min (17,53 m/sec); max. range (with 45 min reserves), 2,032 mls (3 270 km).

Weights: Empty, 4,100 lb (1 860 kg); max. take-off, 7,350 lb (3 334 kg).

Accommodation: Pilot and co-pilot/passenger on flight deck and up to eight passengers in main cabin.

Status: First of six prototypes (including structural and fatigue specimens) flown on 1 January 1981, with second scheduled to join test programme April 1982. All prototypes are being built at the parent company's Reno, Nevada, facility, but the production line is being established at Aldergrove, Belfast, N. Ireland. Certification is expected in April 1983, by which time the first production aircraft are expected to be flying. Orders for Lear Fan 2100s exceeded 260 by the beginning of 1982.

Notes: Radical in both concept and construction, the Lear Fan 2100 is manufactured primarily of composite materials, using graphite/epoxy construction for the fuselage and all surfaces. The turboshafts drive the pusher propeller via a gearbox drive train and fuel is housed by integral wing tanks.

LEARAVIA LEAR FAN 2100

Dimensions: Span, 39 ft 4 in (11,99 m); length, 40 ft 7 in (12,37 m); height, 12 ft 2 in (3,70 m); wing area, 162·9 sq ft (15,13 m²).

LOCKHEED L-100-30 HERCULES

Country of Origin: USA.
Type: Medium/long-range commercial and military freight transport.
Power Plant: Four 4,508 ehp Allison T56-A-15 turboprops.
Performance: Max. cruising speed (at 120,000 lb/54 430 kg), 361 mph (581 km/h) at 20,000 ft (6 100 m); econ. cruise, 325 mph (523 km/h); max. initial climb, 1,900 ft/min (9,65 m/sec); range (max. payload and 45 min. reserves), 2,005 mls (3 226 km), max. fuel and zero payload), 4,833 mls (7 778 km).
Weights: Operational empty, 73,181 lb (33 194 kg); max. normal take-off, 155,000 lb (70 310 kg); max. overload, 175,000 lb (79 380 kg).
Accommodation: Crew of four on flight deck. The freight hold can accommodate seven cargo pallets, or 93 casualty stretchers and six medical attendants in the aeromedical role. Up to 128 troops or 92 fully-equipped paratroops may be accommodated.
Status: The L-100-30 is a stretched version of the basic Hercules utilised by both commercial and military operators, the equivalent "military" version is the C-130H-30, initial deliveries of which (to Indonesia) commenced in 1980. A total of 1,656 Hercules of all types had been ordered by the beginning of 1982, with 1,637 delivered, and more than 65 of these in commercial service.
Notes: The L-100-30 and C-130H-30 features two fuselage plugs totalling 180 in (4,57 m), and 30 of the RAF's Hercules C Mk 1s (equivalent to C-130H) are currently being "stretched" to similar standards as Hercules C Mk 3s.

140

LOCKHEED L-100-30 HERCULES

Dimensions: Span, 132 ft 7 in (40,41 m); length, 112 ft 9 in (34,37 m); height, 38 ft 3 in (11,66 m); wing area, 1,745 sq ft (162,12 m²).

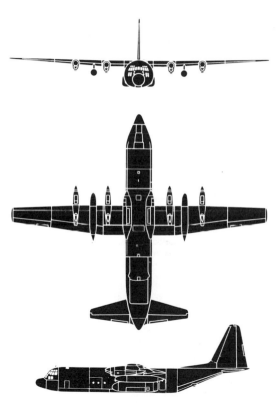

LOCKHEED P-3C ORION

Country of Origin: USA.

Type: Long-range maritime patrol aircraft.

Power Plant: Four 4,910 eshp Allison T56-A-14W turboprops.

Performance: Max. speed (at 105,000 lb/47 625 kg), 473 mph
(761 km/h) at 15,000 ft (4 570 m); cruise, 397 mph (639 km/h)
at 25,000 ft (7 620 m); patrol speed, 230 mph (370 km/h) at
1,500 ft (457 m); loiter endurance (all engines), at 1,500 ft
(457 m), 12·3 hrs, (two engines), 17 hrs; mission radius,
2,530 mls (4 075 km), (with three hours on stations at 1,500 ft/
457 m), 1,933 mls.

Weights: Empty, 61,491 lb (27 890 kg); normal max. take-off,
133,500 lb (60 558 kg); max. overload, 142,000 lb (64 410 kg).

Accommodation: Normal flight crew of 10 including five in
tactical compartment.

Armament: Two Mk 101 depth bombs and four Mk 43, 44 or
46 torpedoes, or eight Mk 54 bombs in weapons bay, and provision for up to 13,713 lb (6 220 kg) external ordnance.

Status: Prototype (YP-3C) flown 8 October 1968, with deliveries to US Navy against planned procurement of 316 P-3Cs
commencing mid-1969. A total of 215 delivered to service by
beginning of 1982, when production was 12 annually for the US
Navy from total production rate (including export) of 16 per
annum. Plans for the final 90 planned for US Navy delivery from
1984 through 1989 were in doubt at the beginning of 1982
when a decision to terminate production in the 1983 Fiscal Year
was anticipated.

Notes: Deliveries of 13 P-3Cs to Netherlands commenced November 1981, and licence manufacture of 42 for Japan following
delivery of three by parent company.

142

LOCKHEED P-3C ORION

Dimensions: Span, 99 ft 8 in (30,37 m); length, 116 ft 10 in (35,61 m); height, 33 ft 8½ in (10,29 m); wing area, 1,300 sq ft (120,77 m²).

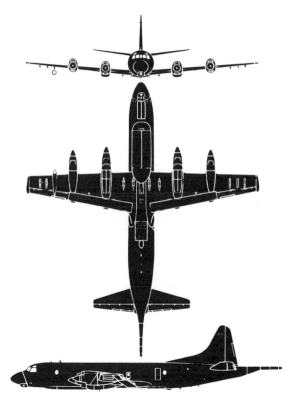

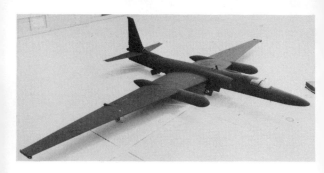

LOCKHEED TR-1A

Country of Origin: USA.

Type: Single-seat high-altitude tactical reconnaissance aircraft.

Power Plant: One 17,000 lb st (7 711 kgp) Pratt & Whitney J75-P-13 turbojet.

Performance: Max. cruising speed, 435 mph (700 km/h) at 70,000 ft (21 355 m); max. operational ceiling, 90,000 ft (27 430 m); max. range 3,000 plus mls (4 830+ km); max. endurance, 12 hrs.

Weights: Approx. max. take-off, 30,000 lb (13 608 kg).

Status: First TR-1A rolled out on 15 July 1981, and delivered to USAF in August against initial requirement for 25 aircraft, including two two-seat TR-1Bs.

Notes: The TR-1A is a derivative of the U-2R and built on the same tooling. Featuring updated sensors and introducing synthetic aperture radar, the TR-1A is intended to orbit 150 miles (240 km) or so behind the battlefield at an altitude of 60,000–70,000 ft (18 290–21 335 m) to provide battlefield reconnaissance information for tactical commanders, its sensors having a range in excess of 300 miles (480 km) and thus providing information relating to movements deep within enemy-held territory without crossing the forward edge of the battle area. The advanced synthetic aperture (side-looking) stand-off radar will produce map-like imagery, the information being passed by data link to ground stations. Modular design permits rapid interchangeability of on-board equipment.

144

LOCKHEED TR-1A

Dimensions: Span, 103 ft 0 in (31,39 m); length, 63 ft 0 in (19,20 m); height, 16 ft 0 in (4,88 m).

LOCKHEED L-1011-500 TRISTAR

Country of Origin: USA.

Type: Long-range commercial transport.

Power Plant: Three 50,000 lb st (22 680 kgp) Rolls-Royce RB211-524B4 turbofans.

Performance: (At 380,000 lb/172 368 kg) Max. cruising speed, 596 mph (959 km/h) at 33,000 ft (10 060 m); econ. cruise, 555 mph (894 km/h) at 33,000 ft (10 060 m); range (with max. payload), 5,270 mls (8 480 km), (with max. fuel), 7,307 mls (11 760 km).

Weights: Operational empty, 245,747 lb (111 471 kg); max. take-off, 496,000 lb (224 982 kg).

Accommodation: Basic flight crew of three and typical mixed-class arrangement for 222 economy (nine-abreast seating) and 24 first-class (six-abreast seating) passengers. Maximum single-class accommodation for 330 passengers.

Status: First L-100-500 flown on 16 October 1978. Total orders for 244 TriStars (all versions) at the beginning of 1982, with 220 delivered and 24 for delivery during course of year, with production to terminate early 1984.

Notes: The L-1011-500 is a shorter-fuselage longer-range derivative of the basic L-1011-1 transcontinental version of the TriStar, a 100-in (254-cm) section having been removed from the fuselage forward of the wing and a 62-in (157,5-cm) section aft of the wing. Versions with the standard fuselage are the L-1011-100 and -200, the latter featuring additional centre-section fuel tankage and similar engines to the -500 (in place of 42,000 lb st/19 050 kgp RB211-22Bs). In abeyance at the time of the late 1981 decision to terminate TriStar production was an extended-range standard-body TriStar with a -500 wing and higher weights.

LOCKHEED L-1011-500 TRISTAR

Dimensions: Span, 164 ft 3½ in (50,07 m); length, 164 ft 2 in (50,04 m); height, 55 ft 4 in (16,87 m); wing area, 3,541 sq ft (328,96 m²).

McDONNELL DOUGLAS DC-9 SUPER 80

Country of Origin: USA.

Type: Short/medium-haul commercial transport.

Power Plant: Two 19,250 lb st (8 730 kgp) Pratt & Whitney JT8D-209 turbofans.

Performance: Max. cruising speed, 574 mph (924 km/h) at 27,000 ft (8 230 m); econ. cruise, 522 mph (840 km/h) at 33,000 ft (10 060 m); long-range cruise, 505 mph (813 km/h) at 35,000 ft (10 670 m); range (with max. payload at econ. cruise), 1,594 mls (2 565 km), (with max. fuel at long-range cruise), 3,280 mls (5 280 km).

Weights: Operational empty, 77,797 lb (35 289 kg); max. take-off, 140,000 lb (63 503 kg).

Accommodation: Flight crew of two and typical mixed-class arrangement for 23 first-class and 137 economy-class passengers, or 155 all-economy or 172 commuter-type arrangements with five-abreast seating.

Status: First Super 80 flown on 18 October 1979, with first customer delivery (to Swissair) on 12 September 1980. Orders for Super 80 totalled 112 at beginning of 1982 of 1,110 for all versions, the 1,000th DC-9 having been delivered (a Super 80 to Swissair) on 3 September 1981.

Notes: The Super 80 is the largest of six members of the DC-9 family and the Super 82, certificated late July 1981 with first customer (Republic) delivery in the following month, is a more powerful and heavier version with 20,850 lb st (9 458 kgp) JT8D-217 engines and a 147,000 lb (66 680 kg) max. take-off weight. Current development studies include the Super 40 utilising the Super 80 wing and cockpit but having a 125·6 ft (38,28 m) fuselage length and either JT8D-200 series or CFM 56-3 engines. A total of 85 DC-9s was delivered during the course of 1981.

148

McDONNELL DOUGLAS DC-9 SUPER 80

Dimensions: Span: 107 ft 10 in (32,85 m); length, 147 ft 10 in (45,08 m); height, 29 ft 4 in (8,93 m); wing area, 1,279 sq ft (118,8 m²).

McDONNELL DOUGLAS DC-10 SERIES 30

Country of Origin: USA.

Type: Medium-range commercial transport.

Power Plant: Three 52,500 lb st (23 814 kgp) General Electric CF6-50C2 turbofans.

Performance: Max. cruising speed (at 396,830 lb/ 180 000 kg), 594 mph (956 km/h) at 31,000 ft (9 450 m); long-range cruise, 540 mph (870 km/h) at 31,000 ft (9 450 m); range (with max. payload), 4,856 mls (7 815 km), (with max. fuel), 6,300 mls (10 140 km).

Weights: Operational empty, 261,459 lb (118 597 kg); max. take-off, 572,000 lb (259 457 kg).

Accommodation: Flight crew of three and typical mixed-class arrangements for 225–270 passengers. Maximum authorised single-class accommodation for 380 passengers.

Status: First DC-10 (Series 10) flown 29 August 1970, with first Series 30 (46th DC-10 built) following on 21 June 1972, this being preceded on 28 February 1972 by first Series 40. Orders (including KC-10A—see pages 158-9) totalled 378 by beginning of 1982, with 363 delivered and 10 scheduled for delivery during course of year.

Notes: The DC-10 Series 30 and 40 have identical fuselages to the DC-10 Series 10 and 15, but whereas these last-mentioned models are intended for domestic operation, the Series 30 and 40 are intercontinental models and differ in power plant, weight and wing details. The intercontinental versions also have three main undercarriage units. The Series 40 has 53,000 lb st (24 040 kgp) Pratt & Whitney JT9D-59A turbofans, but is otherwise similar to the Series 30. The DC-10 Series 15 is a "hot and high" variant combining a Series 10 airframe with a derated version of the Series 30 power plant. The DC-10 Series 30ER is an extended-range version for delivery to Swissair in 1982.

150

McDONNELL DOUGLAS DC-10 SERIES 30

Dimensions: Span, 165 ft 4 in (50,42 m); length, 181 ft 4¾ in (55,29 m); height, 58 ft 0 in (17,68 m); wing area, 3,921·4 sq ft (364,3 m²).

McDONNELL DOUGLAS AV-8B
HARRIER II

Country of Origin: USA (and UK).

Type: Single-seat V/STOL ground attack aircraft.

Power Plant: One 21,180 lb st (9607 kgp) Rolls-Royce F402-RR-406 Pegasus 11-21 E (Mk 105) turbofan.

Performance: Max. speed (clean aircraft), 648 mph (1042 km/h) or Mach 0·85 at sea level, 600 mph (965 km/h) or Mach 0·91 at 36,000 ft (10970 m); tactical radius (HI-LO-HI interdiction mission with seven 500-lb/227-kg Mk 82 bombs and 25-mm cannon), 692 mls (1114 km); ferry range (with four 300 US gal/1135 l auxiliary tanks), 2,833 mls (4560 km).

Weights: Operational empty, 12,750 lb (5783 kg); max. loaded (for STO), 28,750 lb (13041 kg); max. 29,750 lb (13495 kg).

Armament: One 25-mm GAU-12/U five-barrel rotary cannon and up to 9,200 lb (4173 kg) of ordnance on one fuselage centreline and six wing pylons.

Status: First of four FSD (Full Scale Development) AV-8B Harrier IIs flown on 5 November 1981, with remaining three aircraft scheduled for early 1982 completion. The first of 12 pilot production aircraft will be delivered in 1983, and the US Marine Corps as an initial requirement for 257 aircraft (which is likely to be raised to 354 aircraft, including 18 two-seat TAV-8Bs) with initial operational capability being achieved by the first squadron in June 1985. Sixty are to be procured for the RAF (as the Harrier GR Mk 5) with deliveries commencing in 1986.

Notes: The AV-8B is a derivative of the BAe Harrier (see 30–1).

McDONNELL DOUGLAS AV-8B HARRIER II

Dimensions: Span, 30 ft 4 in (9,24 m); length, 46 ft 4 in (14,12 m); height, 11 ft 8 in (3,55 m); wing area, 241 sq ft (22,40 m²).

McDONNELL DOUGLAS F-15C EAGLE

Country of Origin: USA.
Type: Single-seat air superiority fighter.
Power Plant: Two 14,780 lb st (6 705 kgp) dry and 23,904 lb st (10 855 kgp) reheat Pratt & Whitney F100-PW-100 turbofans.
Performance: Max. speed (short-endurance dash), 1,676 mph (2 698 km/h) or Mach 2·54, (sustained), 1,518 mph (2 443 km/h) or Mach 2·3 at 40,000 ft (12 190 m), 922 mph (1 484 km/h) or Mach 1·21 at sea level; max. endurance (internal fuel), 2·9 hrs, (with conformal pallets), 5·25 hrs; ferry range (internal fuel), 1,950 mls (3 138 km), (with conformal pallets), 3,570 mls (5 745 km); service ceiling, 63,000 ft (19 200 m).
Weights: Basic equipped, 28,700 lb (13 018 kg); loaded (full internal fuel and four AIM-7 AAMs), 44,500 lb (20 185 kg); max. take-off, 68,000 lb (30 845 kg).
Armament: One 20-mm M-61A1 rotary cannon plus four AIM-7F Sparrow and four AIM-9L Sidewinder AAMs, or (attack) up to 16,000 lb (7 258 kg) of external ordnance.
Status: First flown on 26 February 1979, the F-15C is the second major single-seat production version of the Eagle. This model, together with its two-seat equivalent, the F-15D, supplanted the F-15A and two-seat F-15B in production from mid-1980 and from the 444th aircraft. Some 670 Eagles had been delivered by the beginning of 1982.
Notes: Forty F-15As have been supplied to Israel, and deliveries of 47 F-15Cs and 15 F-15Ds to Saudi Arabia commenced early 1982. Twelve F-15DJ two-seaters are being supplied to Japan where 84 F-15Js (equivalent to the F-15C) are being manufactured under licence (including eight from knocked-down assemblies) following two supplied by the parent company.

McDONNELL DOUGLAS F-15C EAGLE

Dimensions: Span, 42 ft 9¾ in (13,05 m); length, 63 ft 9 in (19,43 m); height, 18 ft 5½ in (5,63 m); wing area, 608 sq ft (56,50 m²).

McDONNELL DOUGLAS F-18A HORNET

Country of Origin: USA.

Type: Single-seat shipboard fighter.

Power Plant: Two (approx.) 10,600 lb st (4 810 kgp) dry and 15,800 lb st (7 167 kgp) reheat General Electric F404-GE-400 turbofans.

Performance: Max. speed (AAMs on wingtip and fuselage stations), 1,190 mph (1 915 km/h) or Mach 1·8 at 40,000 ft (12 150 m); acceleration from Mach 0·8 to 1·6, 1·8 min at 35,000 ft (10 670 m); initial climb (half fuel and wingtip AAMs), 60,000 ft/min (304,6 m/sec); tactical radius (combat air patrol on internal fuel), 480 mls (770 km), (three external tanks), 740 mls (1 190 km); ferry range, 2,875 mls (4 627 km).

Weights: Empty equipped, 28,000 lb (12 700 kg); loaded (air superiority mission with half fuel and four AAMs), 35,800 lb (16 240 kg); max. take-off, 56,000 lb (25 400 kg).

Armament: One 20-mm M61A-1 rotary cannon and (standard air–air) two AIM-7E/F Sparrow and two AIM-9G/H Sidewinder AAMs, or (attack) up to 17,000 lb (7 711 kg) of ordnance.

Status: First of 11 full-scale development (FSD) Hornets (including two TF-18A two-seaters) flown 18 November 1978. US Navy planning calls for 667 F-18As and TF-18As during 1981–86, a decision of full-scale production of A-18A optimised attack model being in abeyance until autumn 1982. Shore-based multi-role versions ordered by Canada (137 aircraft) and Australia (75 aircraft) with deliveries from late 1982 and late 1984 respectively.

Notes: The F-18A is optimised for the air superiority and combat air patrol roles. Anticipated A-18A production will raise US Hornet procurement to some 1,366 aircraft.

156

McDONNELL DOUGLAS F-18A HORNET

Dimensions: Span, 37 ft 6 in (11,43 m); length, 56 ft 0 in (17,07 m); height, 15 ft 4 in (4,67 m); wing area, 396 sq ft (36,79 m²).

McDONNELL DOUGLAS KC-10A EXTENDER

Country of Origin: USA.

Type: Flight refuelling tanker and military freighter.

Power Plant: Three 52,500 lb st (23 814 kgp) General Electric CF6-50C2 turbofans.

Performance: Max. speed, 620 mph (988 km/h) at 33,000 ft (10 060 m); max. cruise, 595 mph (957 km/h) at 31,000 ft (9 450 m); long-range cruise, 540 mph (870 km/h); typical refuelling mission, 2,200 mls (3 540 km) from base with 200,000 lb (90 720 kg) of fuel and return; max. range (with 170,000 lb/77 112 kg freight), 4,370 mls (7 033 km).

Weights: Operational empty (tanker), 239,747 lb (108 749 kg), (cargo configuration), 243,973 lb (110 660 kg); max. take-off, 590 000 lb (267 624 kg).

Accommodation: Flight crew of five plus provision for six seats for additional crew and four bunks for crew rest. Fourteen additional seats for support personnel may be provided in the forward cabin. Alternatively, a larger area can be provided for 55 more support personnel, with necessary facilities, to increase total accommodation (including flight crew) to 80.

Status: The first KC-10A was flown on 12 July 1980, and 12 had been ordered by the USAF by the beginning of 1982 against long-term USAF plans for 32 aircraft. Follow-on order for further eight (for 1983 delivery) undecided at beginning of 1982.

Notes: The KC-10A is a military tanker/freighter derivative of the commercial DC-10 Series 30 (see pages 150–1) with refuelling boom, boom operator's station, hose and drogue, military avionics and body fuel cells in the lower cargo compartments.

McDONNELL DOUGLAS KC-10A EXTENDER

Dimensions: Span, 165 ft 4 in (50,42 m); length, 182 ft 0 in (55,47 m); height, 58 ft 1 in (17,70 m); wing area, 3,647 sq ft (338,8 m²).

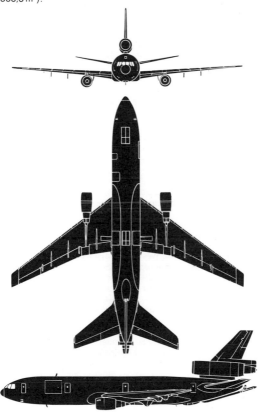

MICROTURBO MICROJET 200B

Country of Origin: France.

Type: Staggered side-by-side two-seat basic trainer.

Power Plant: Two 286 lb st (130 kgp) Microturbo TRS 18-1 turbojets.

Performance: Max. continuous cruising speed, 288 mph (463 km/h); initial climb, 1,738 ft/min (8,83 m/sec); service ceiling, 29,530 ft (9 000 m); range (with 10% reserves), 534 mls (860 km); endurance, 2·25 hrs.

Weights: Empty equipped, 1,462 lb (633 kg); max. take-off, 2,535 lb (1 150 kg).

Status: The prototype Microjet was first flown on 24 June 1980, followed by three pre-series aircraft with first to fly in June 1982.

Notes: The prototype Microjet 200 (see 1981 edition) was subsequently modified to proposed production configuration (as illustrated), the fuselage being lengthened to modify the CG position, the cockpit canopy being recontoured, sweepback being applied to the tail surfaces and the undercarriage retraction and locking systems being revised. Evaluated by both the Armée de l'Air and the Aéronavale, the Microjet is one of a new generation of lightweight jet trainers and is unusual in having a staggered side-by-side seating arrangement for pupil and instructor, the right-hand seat being positioned 21·6 in (55 cm) aft of that on the left which is occupied by the pupil, this layout permitting some small reduction in fuselage cross-section by comparison with conventional side-by-side seating. Whereas the prototype is of wooden construction, the pre-series aircraft are manufactured of composites and metal.

160

MICROTURBO MICROJET 200B

Dimensions: Span, 24 ft 9½ in (7,56 m); length, 21 ft 5⅞ in (6,55 m); wing area, 65·88 sq ft (6,12 m²).

MIKOYAN MIG-23 (FLOGGER)

Country of Origin: USSR.

Type: Single-seat (Flogger-B, E and G) air superiority and (Flogger-F and H) close air support fighter.

Power Plant: One 17,635 lb st (8 000 kgp) and 25,350 lb st (11 500 kgp) reheat Tumansky R-29B turbofan.

Performance: Max. speed (clean aircraft with half fuel), 838 mph (1 350 km/h) or Mach 1·1 at 1,000 ft (305 m), 1,520 mph (2 446 km/h) or Mach 2·3 above 36,090 ft (11 000 m); combat radius (high-altitude air superiority mission with four AAMs), 530 mls (850 km), (with centreline combat tank), 700 mls (1 126 km); ferry range (max. external fuel), 2,100 mls (3 380 km) at (average) 495 mph (795 km/h).

Weights: Normal loaded (clean), 34,170 lb (15 500 kg); max. take-off, 44,312 lb (20 100 kg).

Armament: One 23-mm twin-barrel GSh-23L cannon and (air superiority) two AA-7 Apex semi-active radar-guided and two AA-8 Aphid IR-homing AAMs, or (Flogger-E) four AA-2-2 Advanced Atoll (two IR-homing and two radar-guided) AAMs, or (Flogger-F/H) up to 9,920 lb (4 500 kg) of bombs and missiles.

Status: Aerodynamic prototype flown winter 1966–67, with service debut following 1971. Production rate of 50 monthly at beginning of 1982 (all versions of basic design) when 2,300–2,500 were in Soviet service.

Notes: The MiG-23 has evolved as a family of combat aircraft, the Flogger-E (MiG-23MF) air–air and Flogger-F (MiG-23BM) close support fighters having been widely exported, 70 of a version of the latter (MiG-23BN Flogger-H) being supplied to India, the majority being licence-assembled. Flogger-G is an improved air–air version, and Flogger-D and J (MiG-27) are dedicated tactical strike models (see 1981 edition).

MIKOYAN MIG-23 (FLOGGER)

Dimensions: (Estimated) Span (17 deg sweep), 46 ft 9 in
(14,25 m), (72 deg sweep), 27 ft 6 in (8,38 m); length (including
probe), 55 ft 1½ in (16,80 m); wing area, 293·4 sq ft (27,26 m²).

MIKOYAN MIG-25 (FOXBAT)

Country of Origin: USSR.

Type: Single-seat (Foxbat-A) interceptor fighter and (Foxbat-B and -D) high-altitude reconnaissance aircraft.

Power Plant: Two 20,500 lb st (9 300 kgp) dry and 27,120 lb st (12 300 kgp) reheat Tumansky R-31 turbojets.

Performance: (Foxbat-A) Max. speed (short-period dash with four AAMs), 1,850 mph (2 980 km/h) or Mach 2·8 above 36,000 ft (10 970 m); max. speed at sea level, 650 mph (1 045 km/h) or Mach 0·85; initial climb, 40,950 ft/min (208 m/sec); service ceiling, 80,000 ft (24 385 m); combat radius (including allowance for Mach 2·5 intercept), 250 mls (400 km), (range optimised profile at econ. power), 400 mls (645 km).

Weights: (Foxbat-A) Empty equipped, 44,100 lb (20 000kg); max. take-off, 77,160 lb (35 000 kg).

Armament: Four AA-6 Acrid AAMs (two semi-active radar homing and two IR homing).

Status: The MiG-25 entered service (in Foxbat-A form) in 1970, the photo/ELINT recce version (Foxbat-B) following in 1971 and the optimised ELINT version (Foxbat-D) in 1974.

Notes: The Foxbat-A and tandem two-seat conversion training Foxbat-C have been exported to Algeria, Syria and Libya, and the latter to India, together with Foxbat-C. The -B and -D versions are reportedly capable of Mach 3·2 in clean condition. An advanced tandem two-seat interceptor version of the basic design carrying up to eight AA-9 long-range radar-guided AAMs and possessing full lookdown-shootdown capability was reportedly under development at the beginning of 1982, when it was expected to be deployed operationally within two–three years.

MIKOYAN MIG-25 (FOXBAT)

Dimensions: Span, 45 ft 9 in (13,94 m); length, 73 ft 2 in (22,30 m); height, 18 ft 4½ in (5,60 m); wing area, 602·8 sq ft (56,00 m²).

MITSUBISHI MU-300 DIAMOND I

Country of Origin: Japan.

Type: Light business executive transport.

Power Plant: Two 2,500 lb st (1 134 kgp) Pratt & Whitney (Canada) JT15D-4 turbofans.

Performance: Max. cruising speed, 497 mph (800 km/h) at 30,000 ft (9 145 m); normal cruise, 466 mph (750 km/h); long-range cruise, 426 mph (685 km/h) at 39,000 ft (11 890 m); initial climb, 3,100 ft/min (15,75 m/sec); time to 39,000 ft (11 885 m), 34 min; max. altitude, 41,000 ft (12 495 m); max. range (with four passengers and IFR reserves), 1,484 mls (2 389 km), (VFR reserves), 1,772 mls (2 852 km).

Weights: Empty equipped, 8,845 lb (4 010 kg); max. take-off, 14,430 lb (6 545 kg).

Accommodation: Pilot and co-pilot/passenger on flight deck and six–eight passengers in main cabin.

Status: First of two prototypes flown on 29 August 1978, and first production aircraft flown 21 May 1981, certification being obtained on 6 November 1981, and 120 having been ordered when customer deliveries commenced January 1982.

Notes: The Diamond is manufactured by Mitsubishi at Nagoya, but final assembly from component sets, finishing and equipping is undertaken by the Mitsubishi facility at San Angelo, Texas, from where, after flight testing, the aircraft are sent to distribution centres. Production of the Diamond is scheduled to attain eight monthly during the course of 1982, rising to 10 monthly in 1983. Several versions of the aircraft were under study at the beginning of 1982, including one embodying a modest fuselage stretch.

166

MITSUBISHI MU-300 DIAMOND I

Dimensions: Span, 43 ft 5 in (13,23 m); length, 48 ft 4 in (14,73 m); height, 13 ft 9 in (4,19 m); wing area, 241·4 sq ft (22,43 m²).

NDN-6 FIELDMASTER

Country of Origin: United Kingdom.
Type: Two-seat agricultural aircraft.
Power Plant: One 750 shp Pratt & Whitney (Canada) PT6A-34AG turboprop.
Performance: (Estimated) Max. speed (clean), 188 mph (303 km/h); cruise (75% power), 171 mph (275 km/h); max. initial climb, 1,1200 ft/min (6,1 m/sec); range (max. fuel and no reserves), 1,174 mls (1 889 km).
Weights: (Estimated) Empty equipped, 3,500 lb (1 588 kg); max. take-off, 8,500 lb (3 856 kg).
Status: First flight of NDN-6 prototype took place on 17 December 1981, with certification planned for early 1983, and first production aircraft mid-1983.
Notes: The NDN-6 Fieldmaster is being financed jointly by NDN Aircraft and the British National Research and Development Corporation. The first agricultural aircraft to be designed from the outset to be powered by a turboprop, the NDN-6 will normally be flown as a single-seater, but the cockpit provides accommodation for a second person (eg, mechanic/loader, pupil pilot or fire spotter). The integral hopper/tank (which has a capacity of 581 Imp gal/2 642 l) is part of the primary fuselage structure, carrying the engine bearers at its forward end and the rear fuselage with cockpit aft, a system permitting a smaller cross-section to be utilised for a given capacity. A liquid spray dispersal system is incorporated in a full-span auxiliary aerofoil flap, and all fuel is housed in the outer wing panels for maximum safety.

168

NDN-6 FIELDMASTER

Dimensions: Span, 50 ft 3 in (15,32 m); length, 36 ft 2 in (10,97 m); height, 11 ft 5 in (3,48 m); wing area, 338 sq ft (31,42 m²).

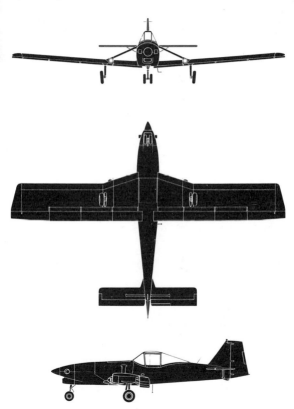

NORTHROP F-5G TIGERSHARK

Country of Origin: USA.
Type: Single-seat multi-role fighter.
Power Plant: One (approx.) 10,600 lb st (4 810 kgp) dry and 16,000 lb st (7 260 kgp) reheat General Electric F404-GE-F1G1 turbofan.
Performance: (Estimated) Max. speed, 1,320 mph (2 124 km/h) or Mach 2·0 above 36,000 ft (10 975 m), 800 mph (1 288 km/h) or Mach 1·05 at sea level; initial climb at combat weight (half fuel and two AAMs), 50,300 ft/min (255,5 m/sec); combat ceiling, 53,700 ft (16 370 m); acceleration from Mach 0·9 to 1·6 at 30,000 ft (9 145 m), 1·33 min; ferry range (max. external fuel), 1,716 mls (2 760 km).
Weights: (F-5G-1) Empty, 11,220 lb (5 089 kg); loaded (clean), 17,240 lb (7 820 kg); max. 26,140 lb (11 857 kg).
Armament: (F-5G-1) Two 20-mm M39 cannon and up to six AIM-9 Sidewinder AAMs.
Status: The first of four flight-test Tigersharks is scheduled to fly in September 1982, with production deliveries of the F-5G-1 commencing July 1983.
Notes: Evolved from the F-5E Tiger II (see 1981 edition), the Tigershark will offer 48% better acceleration, 32% faster climb and 38% higher speed. The initial production model, the F-5G-1, will have essentially similar systems to those of the F-5E, but the more advanced F-5G-2 will be fitted with integrated digital avionics, including a digital flight control system, and a multi-mode coherent pulse-Doppler radar incorporating a continuous-wave illuminator which will permit use of such radar-guided AAMs as the AIM-7E/F Sparrow.

NORTHROP F-5G TIGERSHARK

Dimensions: Span, 26 ft 8 in (8,13 m); length, 46 ft 6⅔ in (14,19 m); height, 13 ft 10¼ in (4,22 m); wing area, 186 sq ft (17,28 m²).

PANAVIA TORNADO F Mк 2

Country of Origin: United Kingdom.
Type: Tandem two-seat interceptor fighter.
Power Plant: Two (approx.) 9,000 lb st (4 082 kgp) dry and
16,000 lb st (7 258 kgp) reheat Turbo-Union RB. 199-34R-04
Mk 101 (Improved) turbofans.
Performance: (Estimated) Max. speed, 920 mph (1 480 km/h)
or Mach 1·2 at sea level, 1,450 mph (2 333 km/h) or Mach 2·2
at 40,000 ft (12 190 m); combat radius (combat air patrol mis-
sion with drop tanks and allowance for 2 hrs plus loiter and
10 min combat), 350–450 mls (560–725 km); ferry range,
2,650 mls (4 265 km).
Weights: Empty equipped (approx.), 25,000 lb (11 340 kg);
max. take-off, 52,000 lb (23 587 kg).
Armament: One 27-mm IWKA-Mauser cannon, two AIM-9L
Sidewinder and four BAe Sky Flash AAMs.
Status: First of three prototype Tornado F Mk 2s flown on 27
October 1979, with production deliveries of 165–185 for the
RAF commencing in 1983.
Notes: The Tornado F Mk 2, which is scheduled to attain initial
operational capability with the RAF in 1984, is a UK-only derivative of the multi-national (UK, Federal Germany and Italy)
multi-role fighter (see 1978 edition) which will attain operational status with the RAF during 1982 as the Tornado GR Mk
1. Retaining 80% commonality with the GR Mk 1, the F Mk 2
features a redesigned nose for the intercept radar and a longer
fuselage which increases internal fuel capacity and permits the
mounting of four Sky Flash missiles on fuselage stations. The
Tornado F Mk 2 places emphasis on range and endurance in
order to mount combat air patrols at considerable distances from
the British coastline.

PANAVIA TORNADO F Mᴋ 2

Dimensions: Span (25 deg), 45 ft 7¼ in (13,90 m), (68 deg), 28 ft 2½ in (8,59 m); length, 59 ft 3 in (18,06 m); wing area, 322·9 sq ft (30,00 m²).

PILATUS PC-7 TURBO TRAINER

Country of Origin: Switzerland.
Type: Tandem two-seat basic trainer.
Power Plant: One (flat-rated) 550 shp Pratt & Whitney (Canada) PT6A-25A turboprop.
Performance: Max. continuous speed, 255 mph (411 km/h) at 10,000 ft (3 050 m); econ. cruise, 230 mph (370 km/h) at 20,000 ft (6 100 m); initial climb, 2,065 ft/min (10,4 m/sec); service ceiling, 31,000 ft (9 450 m); max. range (internal fuel), 650 mls (1 047 km).
Weights: Empty equipped, 2,932 lb (1 330 kg); max. aerobatic, 4,188 lb (1 900 kg); max. take-off, 5,952 lb (2 700 kg).
Armament: (Training or light strike) Six wing hardpoints permit external loads up to max of 2,292 lb (1 040 kg).
Status: First of two PC-7 prototypes flown 12 April 1966, and first production example flown 18 August 1978, with first customer deliveries (to Burma) following early 1979. Total sales exceeded 290 aircraft at beginning of 1982, in which year planned production is approximately six–seven monthly, with 82 aircraft scheduled for completion.
Notes: Derived from the piston-engined P-3 basic trainer, the PC-7 has been selected for service by 10 air arms as follows: Abu Dhabi (14), Bolivia (36), Burma (17), Chile (10), Guatemala (12), Iraq (52), Malaysia (44), Mexico (55), Switzerland (40) and an unspecified African country (12). The PC-7 combines the basic training and close air support roles in some air forces (e.g., Mexico), and is capable of carrying a variety of bombs, rocket and machine gun pods and other ordnance externally, provision being made for weapons aiming sight, etc.

PILATUS PC-7 TURBO TRAINER

Dimensions: Span, 34 ft 1½ in (10,40 m); length, 31 ft 11⅞ in (9,75 m); height, 10 ft 6½ in (3,21 m); wing area, 178·68 sq ft (16,60 m²).

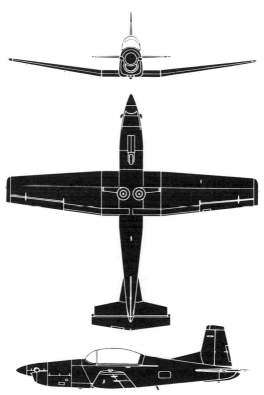

PILATUS BRITTEN-NORMAN
BN-2T TURBINE ISLANDER

Country of Origin: United Kingdom.

Type: Light utility transport.

Power Plant: Two (flat-rated) 320 shp Allison 250-B17C turboprops.

Performance: Max. cruising speed, 180 mph (290 km/h) at sea level; cruise (75% power), 161 mph (259 km/h) at 5,000 ft (1 525 m); initial climb, 1,200 ft/min (6,1 m/sec); range (VFR and no reserves), 498 mls (802 km), (with two 29 Imp gal/132 1 underwing tanks), 782 mls (1 258 km), (with two 50 Imp gal/227 1 tanks), 985 mls (1 585 km).

Weights: Empty equipped, 3,745 lb (1 699 kg); max. take-off, 6,600 lb (2 994 kg).

Accommodation: Flight crew of one or two and up to nine passengers (one beside pilot and four double seats).

Status: The BN-2T prototype was flown on 2 August 1980, and initial production aircraft (conversion of a BN-2B) was flown mid-1981. At beginning of 1982, anticipated production rate of 4-5 Islanders monthly during course of year was expected to include one BN-2T.

Notes: The BN-2T is essentially a re-engined BN-2B, which, with either the 260 hp Avco Lycoming 0-540-E4C5 (BN-2B-26/27) or 300 hp 10-540-K1B5 "flat-six" engines, remained the principal production models of the Islander at the beginning of 1982. All airframes are manufactured in Romania and supplied to the Pilatus Britten-Norman Bembridge facility for equipping and finishing.

PILATUS BRITTEN-NORMAN BN-2T TURBINE ISLANDER

Dimensions: Span, 49 ft 0 in (14,94 m); length, 35 ft 7¾ in (10,86 m); height, 13 ft 9 in (4,19 m); wing area, 325 sq ft (30,19 m²).

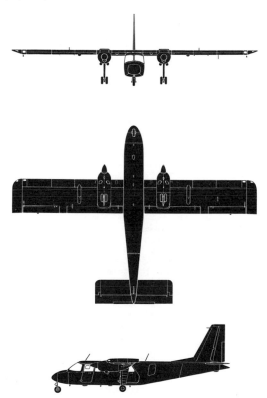

PIPER PA-42 CHEYENNE III

Country of Origin: USA.

Type: Light business executive transport.

Power Plant: Two (flat-rated) 720 shp Pratt & Whitney (Canada) PT6A-41 turboprops.

Performance: Max. cruising speed, 334 mph (537 km/h) at 20,000 ft (6 095 m), 317 mph (511 km/h) at 30,000 ft (9 145 m); initial climb (at 11,200 lb/5 080 kg), 2,236 ft/min (11,36 m/sec); service ceiling, 32,000 ft (9 755 m); range (at max. cruise with 45 min reserves), 1,611 mls (2 593 km) at 20,000 ft (6 095 m), 2,300 mls (3 704 km) at 33,000 ft (10 060 m), (at long-range cruise), 2,054 mls (3 306 km) at 20,000 ft (6 095 m), 2,577 mls (4 148 km) at 33,000 ft (10 060 m).

Weights: Operational empty, 6,389 lb (2 898 kg); max. take-off, 11,200 lb (5 080 kg).

Accommodation: Flight crew of one or two on separate flight deck and various optional arrangements for from six to nine passengers in cabin, with baggage compartments in nose, aft cabin and engine nacelle extensions.

Status: The production prototype Cheyenne III was flown on 18 May 1979, certification following on 18 December, and customer deliveries commencing 30 June 1980. More than 60 delivered by the beginning of 1982.

Notes: Flight development of the Cheyenne III commenced in 1977, but this model subsequently underwent considerable revision before achieving production configuration. The 1982 model offers optional nacelle fuel tanks adding 170 US gal (643,5 l) of usable fuel and other refinements.

PIPER PA-42 CHEYENNE III

Dimensions: Span, 47 ft 8⅛ in (14,53 m); length, 43 ft 4¾ in (13,23 m); height, 14 ft 9 in (4,50 m); wing area, 293 sq ft (27,20 m²).

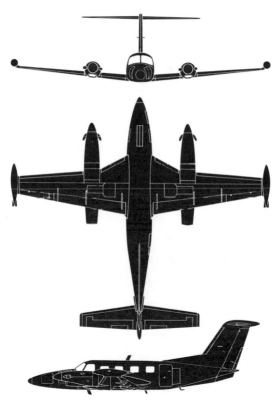

PIPER PA-28R-300 XBT PILLAN

Country of Origin: USA.

Type: Tandem two-seat primary/basic trainer.

Power Plant: One 300 bhp Avco Lycoming AEIO-540 six-cylinder horizontally-opposed engine.

Performance: Max. speed, 199 mph (321 km/h) at sea level; cruise (75% power), 191 mph (308 km/h) at 7,600 ft (2 320 m), (65% power), 185 mph (298 km/h) at 11,300 ft (3 450 m), (55% power), 176 mph (283 km/h) at 15,200 ft (4 630 m); initial climb, 1,240 ft/min (6,3 m/sec); time to 9,840 ft (3 000 m), 11·0 min; range (with 45 min reserves), 702 mls (1 130 km) at 75% power at 8,200 ft (2 500 m); max. range, 956 mls (1 538 km) at 13,100 ft (4 000 m).

Weights: Max. take-off, 2,900 lb (1 315 kg).

Status: First prototype flown spring 1981, with second following in autumn 1981 for service evaluation in Chile. Subject to satisfactory completion of trials that were continuing at the beginning of 1982, the Pillan is to be supplied in CKD (Completely Knocked Down) form to the Maintenance Wing of the Chilean Air Force at El Bosque by the parent company for licence assembly. The Chilean Air Force has a requirement for up to 100 Pillan trainers with deliveries commencing late 1982.

Notes: The Pillan aerobatic trainer has been developed by Piper to meet the requirements of a Chilean Air Force specification, the name Pillan meaning Demon in the dialect of the Araucanos Indians of Southern Chile. The Pillan utilises the basic wing of the PA-28 Cherokee, the Chilean Air Force's Maintenance Wing having assembled a number of PA-28 Cherokee Dakotas from CKD kits, and apart from the centre fuselage, all elements of the Pillan are based on existing components.

180

PIPER PA-28R-300XBT PILLAN

Dimensions: Span, 28 ft 11 in (8,81 m); length, 26 ft 1 in (7,97 m); height, 7 ft 8½ in (2,34 m); wing area, 147 sq ft (13,64 m²).

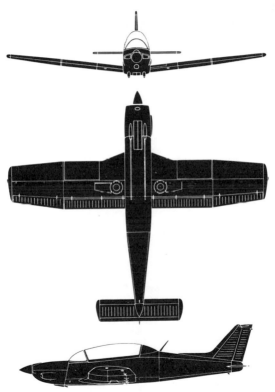

PIPER T-1040

Country of Origin: USA.

Type: Light short-haul regional transport.

Power Plant: Two 500 shp Pratt & Whitney (Canada) PT6A-11 turboprops.

Performance: Max. speed, 280 mph (450 km/h); max. cruise, 234 mph (376 km/h) at sea level, 274 mph (440 km/h) at 11,000 ft (3 355 m); range (with max. payload), 518 mls (834 km) at 274 mph (440 km/h) at 10,000 ft (3 050 m), (with nine passengers and baggage), 772 mls (1 242 km).

Weights: Empty, 5,230 lb (2 372 kg); max. take-off, 9,000 lb (4 082 kg).

Accommodation: Eleven individual seats in cabin including one or two pilots' seats with full dual control. Baggage lockers in nose, rear of cabin and engine nacelle extensions with max. capacity of 25·5 cu ft (0,72 m³).

Status: Prototype of T-1040 flown for the first time on 17 July 1981, with initial customer deliveries scheduled to commence April 1982.

Notes: An optimised commuterliner utilising the basic wing, nose and tail of the PA-31T Cheyenne mated with the fuselage of the PA-31-350 Chieftain, the T-1040 is being manufactured in parallel with the T-1020 as the first two Piper aircraft evolved specifically for airline use. The T-1020, affording similar capacity to that of the T-1040, is powered by 350 hp Avco Lycoming TIO/LTIO-540-J2BD six-cylinder horizontally-opposed piston engines, and is, in fact, the Chieftain with an optimised interior. Both the T-1020 and T-1040 are aimed at the lower end of the commuter market, and embody experience gained by small airlines operating more than 500 PA-31-350 Chieftains on scheduled commuter services.

PIPER T-1040

Dimensions: Span, 41 ft 1 in (12,52 m); length, 36 ft 8 in (11,18 m); height, 12 ft 9 in (3,89 m); wing area, 229 sq ft (21,27 m²).

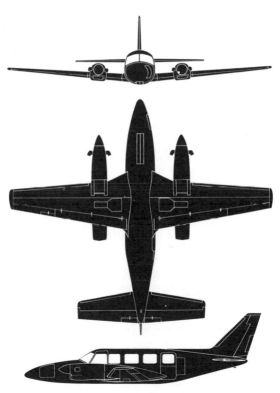

SAAB (JA) 37 VIGGEN

Country of Origin: Sweden.

Type: Single-seat all-weather interceptor fighter with secondary strike capability.

Power Plant: One 16,200 lb st (7 350 kgp) dry and 28,110 lb st (12 750 kgp) reheat Volvo Flygmotor RM 8B turbofan.

Performance: Max. speed (with four AAMs), 838 mph (1 350 km/h) of Mach 1·1 at sea level, 1,255-1,365 mph (2 020–2 195 km/h) or Mach 1·9–2·1 at 36,090 ft (11 000 m); time (from brakes off) to 32,810 ft (10 000 m), 1·4 min; tactical radius (Mach 2·0 intercept mission), 250 mls (400 km), (counterair mission, HI-LO-HI profile with centreline drop tank and 3,000 lb/1 360 kg external load), 650 mls (1 046 km), (LO-LO-LO), 300 mls (480 km).

Weights: Empty (approx.), 26,895 lb (12 200 kg); combat (cannon armament and half fuel), 33,070 lb (15 000 kg), (with four AAMs), 37,040 lb (16 800 kg); max. take-off, 49,600 lb (22 500 kg).

Armament: One 30-mm Oerlikon KCA cannon and (intercept) two Rb 71 Sky Flash and two (or four) Rb 24 Sidewinder AAMs, or (interdiction) 13,227 lb (6 000 kg) of external ordnance.

Status: First of four JA 37 prototypes (modified from AJ 37 airframes) flown June 1974, with fifth and definitive prototype flown 15 December 1975. First production JA 37 flown on 4 November 1977. Total of 149 JA 37s (of 329 Viggens of all types) being produced for Swedish Air Force, with final deliveries scheduled for 1985.

Notes: The JA 37 is an optimised interceptor derivative of the AJ 37 attack aircraft (see 1973 edition).

SAAB (JA) 37 VIGGEN

Dimensions: Span, 34 ft 9¼ in (10,60 m); length (excluding probe) 50 ft 8¼ in (15,45 m); height, 19 ft 4¼ in (5,90 m); wing area (including foreplanes), 561·88 sq ft (52,20 m²).

SAAB-FAIRCHILD 340

Countries of Origin: Sweden and USA.

Type: Short-haul regional transport.

Power Plant: Two 1,675 shp General Electric CT7-5A or 1,645 shp CT7-7 turboprops.

Performance: (Estimated) Max. cruising speed, 299 mph (482 km/h) at 20,000 ft (6 095 m); long-range cruise, 243 mph (391 km/h); range (34 passengers and IFR reserves), 944 mls (1 520 km), (22 passengers), 1 886 mls (3 035 km).

Weights: Operational empty, 14,700 lb (6 668 kg); Max. take-off, 25,000 lb (11 340 kg).

Accommodation: Flight crew of two and standard commuter arrangement for 34 passengers three-abreast. Corporate executive transport arrangement for 14 passengers, with optional arrangements for 12 to 24 passengers.

Status: First prototype scheduled to fly December 1982, with deliveries expected to commence in 1984, 24 aircraft being produced during the course of that year. Planned production rate of six monthly by mid-1986.

Notes: The SF 340 is being jointly developed, built and marketed by Saab-Scania in Sweden and Fairchild Swearingen in the USA, the former being responsible for fuselage construction, final assembly and flight testing up to certification, and Fairchild Industries being responsible for manufacture of the wings, tail surfaces and engine nacelles. Commuterliner and executive transport versions are to be built in parallel, the former with CT7-5A and the latter with CT7-7 engines.

186

SAAB-FAIRCHILD 340

Dimensions: Span, 70 ft 4 in (21,44 m); length, 64 ft 6 in (19,67 m); height, 22 ft 6 in (6,87 m).

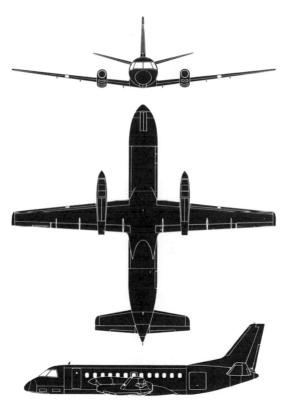

SEPECAT JAGUAR INTERNATIONAL

Countries of Origin: United Kingdom and France.
Type: Single-seat tactical strike fighter.
Power Plant: Two 5,520 lb st (2 504 kgp) dry and 8,400 lb st (3 811 kgp) reheat Rolls-Royce/Turboméca RT172-58 Adour 811 turbofans.
Performance: Max. speed, 820 mph (1 320 km/h) or Mach 1·1 at sea level, 1,057 mph (1 700 km/h) or Mach 1·6 at 32,810 ft (10 000 m); range (LO-LO-LO mission profile with external fuel), 564 mls (907 km), (HI-LO-HI), 875 mls (1 408 km); ferry range, 2,190 mls (3 524 km).
Weights: Typical empty, 15,432 lb (7 000 kg); normal loaded (clean aircraft), 10,000 lb (4 536 kg); max. take-off, 34,000 lb (15 422 kg).
Armament: Two 30-mm Aden cannon and up to 10,000 lb (4 536 kg) of ordnance on five external stations. Provision for two Matra Magic AAMs on overwing stations or AIM-9P Sidewinder AAMs on underwing stations.
Status: The Jaguar International is the current production (by British Aerospace) version of the basic Jaguar of which the first of eight prototypes was flown on 8 September 1968. A total of 202 (including 37 two-seaters) of the basic version was delivered to the RAF and 200 (including 40 two-seaters) to the French Air Force.
Notes: Although manufacture of the Jaguar International is shared between British Aerospace and Dassault-Breguet, the former is responsible for all assembly, having supplied 12 to each of Ecuador and Oman, a further 12 being on order for the latter for 1983 delivery, and 40 (including five two-seaters) to India. At the beginning of 1982, a further 45 were being supplied to India in knocked-down component kit form for assembly by Hindustan Aeronautics.

SEPECAT JAGUAR INTERNATIONAL

Dimensions: Span, 28 ft 6 in (8,69 m); length, 50 ft 11 in (15,52 m); height, 16 ft 0½ in (4,89 m); wing area, 280·3 sq ft (24,18 m²).

SHORTS 330

Country of Origin: United Kingdom.

Type: Short-range regional and utility transport.

Power Plant: Two (330-100) 1,173 shp Pratt & Whitney (Canada) PT6A-45B or (330-200) 1,198 shp PT6A-45R turboprops.

Performance: (330-200) High-speed cruise, 219 mph (352 km/h) at 10,000 ft (3 050 m); long-range cruise, 181 mph (291 km/h) at 6,000 ft (1 830 m); initial climb, 1,180 ft/min (5,99 m/sec); range (with max. fuel), 775 mls (1 247 km), (with max. payload), 320 mls (515 km).

Weights: Operational empty, 15,000 lb (6 805 kg); max. take-off, 22,900 lb (10 387 kg).

Accommodation: Flight crew of two, one flight attendant and normal arrangement for 30 passengers in 10 rows three-abreast and 1,000 lb (455 kg) of baggage.

Status: Engineering prototype of the Shorts 330 flown on 22 August 1974, with production prototype following on 8 July 1975. First production aircraft flown on 15 December 1975, with customer deliveries commencing mid-1976. At the beginning of 1982, orders totalled 105 aircraft with approximately 80 aircraft delivered.

Notes: The Shorts 330-200, announced mid-1981, incorporates a number of product improvements over the basic 330, featuring as standard several items previously listed as options and differing primarily in having similar engines to those of the Shorts 360 (see pages 192–3). Whereas the PT6A-45B of the basic 330 has a water–methanol power restoration system, the PT6A-45R incorporates a reserve power system allowing an increase in max. take-off weight as well as a saving in weight by elimination of the water–methanol system.

SHORTS 330

Dimensions: Span, 74 ft 8 in (22,76 m); length, 58 ft 0½ in (17,69 m); height, 16 ft 3 in (4,95 m); wing area, 453 sq ft (42,10 m²).

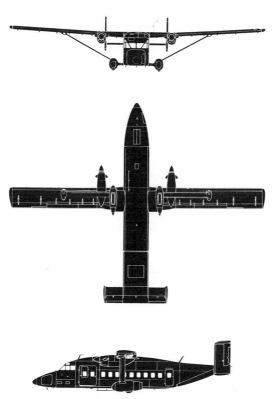

SHORTS 360

Country of Origin: United Kingdom.

Type: Short-range regional and utility transport.

Power Plant: Two flat-rated 1,198 shp Pratt & Whitney (Canada) PT6A-45R turboprops.

Performance: High-speed cruise, 243 mph (391 km/h) at 10,000 ft (3 050 m); range at max. cruise (with max payload and allowances for 100-mile/160-km diversion and 45 min hold), 265 mls (426 km), (with max fuel and same reserves), 655 mls (1 054 km).

Weights: Operational empty, 16,490 lb (7 480 kg); max. take-off, 25,700 lb (11 657 kg).

Accommodation: Flight crew of two, one flight attendant and standard arrangement for 36 passengers in 11 rows three-abreast. Baggage compartment in nose and aft of cabin.

Status: The prototype Shorts 360 flew for the first time on 1 June 1981, and 21 had been ordered by the beginning of 1982, with first customer deliveries scheduled for the last quarter of the year.

Notes: The Shorts 360 is a growth version of the 330 (see pages 190–1) differing from its progenitor primarily in having a 3-ft (91-cm) cabin stretch ahead of the wing and an entirely redesigned rear fuselage and tail assembly. The fuselage lengthening permits passenger capacity to be increased by two rows of three seats, and the lower aerodynamic drag by comparison with the earlier aircraft contributes to a higher performance.

SHORTS 360

Dimensions: Span, 74 ft 8 in (22,75 m); length, 70 ft 6 in (21,49 m); height, 22 ft 7 in (6,88 m); wing area, 453 sq ft (42,08 m²).

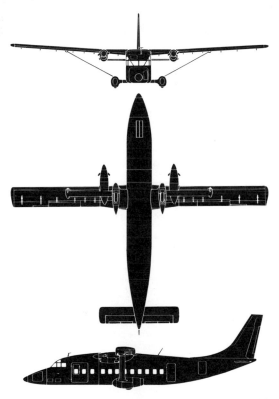

SIAI MARCHETTI S.211

Country of Origin: Italy.

Type: Tandem two-seat basic trainer.

Power Plant: One 2,500 lb st (1 134 kgp) Pratt & Whitney (Canada) JT15D-4C turbofan.

Performance: Max. speed, 449 mph (723 km/h) at 25,000 ft (7 620 m); max. cruise, 437 mph (704 km/h) at 25,000 ft (7 620 m); max. climb, 4,950 ft/min (25,15 m/sec); service ceiling, 42,000 ft (12 800 m); max. range (internal fuel and 30 min reserves), 1,187 mls (1 910 km) at 30,000 ft (9 145 m), (with two external tanks), 1,673 mls (2 693 km).

Weights: Empty, 3,186 lb (1 445 kg); normal loaded, 5,070 lb (2 300 kg); max. take-off, 6,173 lb (2 800 kg).

Armament: (For armament training and light strike) Four wing hardpoints each stressed for loads up to 660 lb (300 kg) inboard and 330 lb (150 kg) outboard, with maximum external load of 1,320 lb (600 kg).

Status: First of two prototypes flown on 10 April 1981, with first production deliveries scheduled for second half of 1982.

Notes: Developed as a private venture and as an attempt to arrest the upward spiralling cost of training a military pilot, the S.211 is substantially less than half the empty weight of other new-generation conventional jet trainers (e.g., Alpha Jet and Hawk) and barely heavier than such turboprop trainers as the Pilatus PC-7. It is nevertheless a comparatively sophisticated design with a supercritical wing section and the ability to accommodate a wide range of avionics. Although no orders for the S.211 had been announced at the time of closing for press, at least a dozen air forces are claimed to be showing strong interest in this trainer.

SIAI MARCHETTI S.211

Dimensions: Span, 26 ft 2⅞ in (8,00 m); length, 30 ft 5½ in (9,28 m); height, 12 ft 2¾ in (3,73 m); wing area, 135·63 sq ft (12·60 m²).

SIAI MARCHETTI SF.260-TP

Country of Origin: Italy.

Type: Side-by-side two-seat primary/basic trainer.

Power Plant: One 350 shp Allison 250B-17C turboprop.

Performance: Max. speed, 237 mph (382 km/h) at sea level; max. continuous cruise, 230 mph (371 km/h) at 10,000 ft (3 050 m); initial climb, 2,170 ft/min (11 m/sec); service ceiling, 28,000 ft (8 535 m); max. range (with 30 min reserves), 590 mls (950 km).

Weights: Empty equipped, 1,753 lb (795 kg); max. take-off, 2,645 lb (1 200 kg).

Status: Prototype SF.260-TP flown for the first time July 1980, with first production deliveries planned for mid 1982.

Notes: The SF.260-TP is a turboprop-powered derivative of the highly-successful piston-engined SF.260 of which more than 450 have been supplied to military customers and in excess of 100 to civil customers. The SF.260 has been in continuous production since 1966, and the SF.260-TP has been developed to overcome the difficulty in obtaining Avgas in certain areas of the world. The turboprop-powered trainer adheres closely structurally to its piston-engined predecessor and few airframe modifications have taken place aft of the firewall. Siai Marchetti proposes to offer kits for the conversion of SF.260s to SF.260TP standards as well as producing the turboprop-powered trainer as a new aircraft. More than a dozen air forces currently use the SF.260, the largest customer being Libya (which procured a total of 240).

SIAI MARCHETTI SF.260-TP

Dimensions: Span, 27 ft 4¾ in (8,35 m); length, 24 ft 3½ in (7,40 m); height, 7 ft 10⅞ in (2,41 m); wing area, 108·72 sq ft (10,10 m²).

SIAI MARCHETTI SF.600-TP

Country of Origin: Italy.
Type: Light utility transport.
Power Plant: Two 420 shp Allison 250B-17C turboprops.
Performance: Max. speed, 196 mph (315 km/h) at 10,000 ft (3 050 m); max. cruise, 181 mph (291 km/h); initial climb, 1,520 ft/min (7,72 m/sec); service ceiling, 21,980 ft (6 700 m); max. range (internal fuel), 982 mls (1 580 km); ferry range (with external tanks), 1,398 mls (2 250 km).
Weights: Empty, 3,968 lb (1 800 kg); normal loaded, 7,275 lb (3 300 kg); max. take-off, 8,157 lb (3 700 kg).
Status: Prototype of SF.600-TP first flown on 8 April 1981 (as rework of original piston-engined SF.600). Pre-series of 20 to be built with deliveries from early 1983 at two monthly.
Notes: Derived from the prototype SF.600 Canguro utility transport powered by two 250 hp Avco Lycoming T10-540-J piston engines, the SF.600-TP features turboprop power and a wing of increased span and area. Capable of operating from semi-prepared strips, the SF.600-TP is being offered in a number of versions, including a freighter with a simple, manually-operated swing tail permitting bulky items of freight to be loaded straight into the boxlike hold. As a third-level light transport, the SF.600-TP will accommodate six passengers in individual seats, three more on a bench-type seat across the rear cabin bulkhead and a tenth passenger seated beside the pilot. Other versions include a palletised cargo model fitted with rollers to ease pallet handling and an aeromedical version for four stretcher casualties and two medical attendants.

198

SIAI MARCHETTI SF.600-TP

Dimensions: Span, 49 ft 2½ in (15,00 m); length, 39 ft 10½ in (12,15 m); height, 15 ft 1 in (4,60 m); wing area, 258·34 sq ft (24,00 m²).

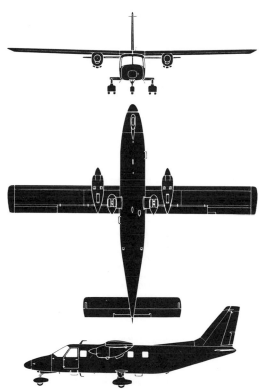

SUKHOI SU-17 (FITTER)

Country of Origin: USSR.

Type: Single-seat ground attack and counterair fighter.

Power Plant: One 17,195 lb st (7 800 kgp) dry and 24,700 lb st (11 200 kgp) reheat Lyulka AL-21F turbojet.

Performance: (Estimated for Fitter-H) Max. speed (short-endurance dash, clean aircraft), 1,430 mph (2 300 km/h) or Mach 2·17 at 39,370 ft (12 000 m), (sustained), 808 mph (1 300 km/h) or Mach 1·06 at sea level, 1,190 mph (1 915 km/h) or Mach 1·8 at 39,370 ft (12 000 m); combat radius (drop tanks on outboard wing pylons and 4,410-lb/2 000-kg external ordnance, LO-LO-LO mission profile), 320 mls (515 km), (HI-LO-HI), 530 mls (853 km).

Weights: (Estimated for Fitter-H) Max. take-off, 39,022 lb (17 700 kg).

Armament: Two 30-mm NR-30 cannon and max. external load of 7,716 lb (3 500 kg) of ordnance.

Status: Variable-geometry derivative of fixed-geometry Su-7 (Fitter-A) first flown as technology demonstrator in 1966 as S-221 (Fitter-B). Initial series Su-17 (Fitter-C) entered Soviet service in 1971, with upgraded model (Fitter-D) following in 1976, and more extensively revised version (Fitter-H) in 1979.

Notes: The Fitter-H (illustrated above) has evolved as the result of an unusual incremental development programme, an equivalent tandem two-seat conversion trainer being the Fitter-G, and a Tumansky R-29B-engined export version of the single-seater, the Fitter-J (illustrated opposite), being referred to as the Su-22. An earlier similarly-engined model, also referred to as Su-22, is Fitter-F, an export version of Fitter-D which introduced terrain-following radar and a laser-marked target seeker.

SUKHOI SU-17 (FITTER)

Dimensions: (Estimated) Span (28 deg), 45 ft 0 in (13,70 m), (68 deg), 32 ft 6 in (9,90 m); length (including probe), 58 ft 3 in (17,75 m); height, 15 ft 5 in (4,70 m); wing area, 410 sq ft (38,00 m²).

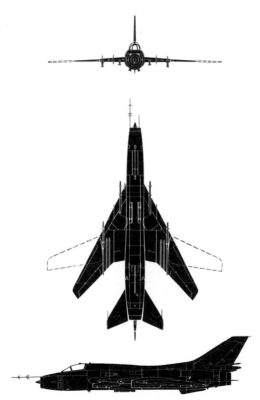

SUKHOI SU-24 (FENCER)

Country of Origin: USSR.

Type: Deep penetration, interdictor and strike aircraft.

Power Plant: Two 17,635 lb st (8 000 kgp) dry and 25,350 lb st (11 500 kgp) reheat Tumansky R-29 turbofans.

Performance: (Estimated) Max. speed (clean), 915 mph (1 470 km/h) or Mach 1·2 at sea level, 1,520 mph (2 446 km/h) or Mach 2·3 above 36,000 ft (11 000 m), (with two combat tanks and full internal fuel), 685 mph (1 102 km/h) or Mach 0·95 at sea level, 1,254 mph (2 018 km/h) or Mach 1·9 above 36,000 ft (11 000 m); tactical radius (HI-LO-HI mission profile with combat tanks and 4,400 lb/2 000 kg ordnance), 1,050 mls (1 690 km), (LO-LO-LO mission profile with same stores), 345 mls (555 km); max. climb rate (half fuel), 35,000 ft/min (178 m/sec).

Weights: (Estimated) Empty equipped, 41,890 lb (19 000 kg); max. take-off, 87 080 lb (39,500 kg).

Armament: One 23-mm six-barrel rotary cannon and one 30-mm cannon, and (short-range interdiction) up to 22 220-lb (100-kg) or 551-lb (250-kg) bombs, or 16 1,102-lb (500-kg) bombs. Alternative loads include AS-7 Kerry ASMs, AS-9, AS-11 or AS-12 anti-radiation ASMs and AS-10 electro-optical ASMs, plus AA-2-2 Advanced Atoll or AA-8 Aphid AAMs for self defence.

Status: Prototype believed to have flown in 1970, with initial operational status achieved late 1974. Production of 8-9 monthly at beginning of 1982, when 550–600 were in Soviet service.

Notes: The first Soviet aircraft designed from the outset for interdiction and counterair missions, the Su-24 carries pilot and weapon systems operator side by side.

SUKHOI SU-24 (FENCER)

Dimensions: (Estimated) Span (16 deg sweep), 56 ft 6 in (17,25 m), (68 deg sweep), 33 ft 9 in (10,30 m); length (excluding probe), 65 ft 6 in (20,00 m); height, 18 ft 0 in (5,50 m); wing area, 452 sq ft (42,00 m²).

TRANSALL C. 160

Countries of Origin: France and Federal Germany.
Type: Medium-range tactical transport.
Power Plant: Two 6,100 eshp Rolls-Royce/SNECMA Tyne RTy 20 Mk 22 turboprops.
Performance: Max. speed, 319 mph (513 km/h) at 16,000 ft (4 875 m); max. continuous cruise, 310 mph (499 km/h) at 20,000 ft (6 100 m); econ. cruise, 282 mph (454 km/h); initial climb (at 108,355 lb/49 150 kg), 1,360 ft/min (6,9 m/sec); max. range (with max. payload and 5% plus 30 min reserves), 1,150 mls (1 850 km), (with max. fuel and 17,637 lb/8 000 kg payload), 5,500 mls (8 854 km).
Weights: Operational empty, 61,728 lb (28 000 kg); max. take-off, 112,434 lb (51 000 kg).
Accommodation: Flight crew of three and up to 35,273 lb (16 000 kg) of freight, or 62–88 paratroops, max. of 93 fully-equipped troops, or up to 63 casualty stretchers and four medical attendants.
Status: First of second (relaunched) production series flown on 9 April 1981, and production orders at beginning of 1982 called for 25 for the Armée de l'Air and three for the Indonesian government, production rate being one per month, with seven flown by January 1982. The first prototype Transall was flown on 25 February 1963, and two further prototypes and 179 production had been built when manufacture terminated in 1972.
Notes: Manufacturing programme shared between France (Aérospatiale) and Germany (MBB/VFW). New series embodies various modifications, the flight crew having been reduced, internal fuel increased and provision made for flight refuelling.

TRANSALL C. 160

Dimensions: Span, 131 ft 3 in (40,00 m); length (excluding refuelling probe), 106 ft 3½ in (32,40 m); height, 38 ft 2½ in (11,65 m); wing area, 1723·3 sq ft (160,10 m²).

TUPOLEV (BACKFIRE-B)

Country of Origin: USSR.

Type: Medium-range strategic bomber and maritime strike/reconnaissance aircraft.

Power Plant: Two (estimated) 33,070 lb st (15 000 kgp) dry and 46,300 lb st (21 000 kgp) reheat Kuznetsov turbofans.

Performance: (Estimated) Max. speed (short-period dash), 1,265 mph (2 036 km/h) or Mach 1·91 at 39,370 ft (12 000 m), (sustained), 1,056 mph (1 700 km/h) or Mach 1·6 at 39,370 ft (12 000 m), 685 mph (1 100 km/h) or Mach 0·9 at sea level; combat radius (unrefuelled with single AS-4 ASM and high-altitude subsonic mission profile), 2,610 mls (4 200 km); max. unrefuelled combat range (with 12,345 lb/5 600 kg internal ordance), 5,560 mls (8 950 km).

Weights: (Estimated) Max. take-off, 260,000 lb (118 000 kg).

Armament: Remotely-controlled tail barbette containing twin 23-mm NR-23 cannon. Internal load of free-falling weapons up to 12,345 lb (5 600 kg) or one AS-4 Kitchen inertially-guided stand-off missile.

Status: Flight testing of initial prototype commenced late 1969, with pre-production series of up to 12 aircraft following in 1972-73. Initial version (Backfire-A) was built in only small numbers. Initial operational capability attained with Backfire-B 1975-76, production rate of 30 annually being attained in 1977 and remaining constant at beginning of 1982, when 75-80 were in service with Soviet Long-range Avitation and a similar quantity with the Soviet Naval Air Force.

TUPOLEV (BACKFIRE-B)

Dimensions: (Estimated) Span (20 deg), 115 ft 0 in (35,00 m), (55 deg) 92 ft 0 in (28,00 m); length, 138 ft 0 in (42,00 m); height, 29 ft 6 in (9,00 m); wing area, 1,830 sq ft (170,00 m²).

TUPOLEV TU-134B-1 (CRUSTY)

Country of Origin: USSR.

Type: Short/medium-haul commercial transport.

Power Plant: Two 14,990 lb st (6 800 kgp) Soloviev D-30 Srs III turbofans.

Performance: Max. cruising speed, 550 mph (885 km/h) at 32,810 ft (10 000 m); econ. cruise, 528 mph (850 km/h); long-range cruise, 497 mph (800 km/h); service ceiling, 39,040 ft (11 900 m); range (with max. payload), 1,174 mls (1 890 km), (with max. fuel), 1,876 mls (3 020 km).

Weights: Operational empty, 64,770 lb (29 380 kg); max. take-off, 104,718 lb (47 500 kg).

Accommodation: Normal flight crew of three with alternative arrangements for 84 or 90 passengers in single-class four-abreast seating. A 100-seat high-density version is under development.

Status: The Tu-134B-1 commenced flight test mid-1981 and will enter Aeroflot service late 1982. Production was initiated (in parallel with Tu-134A-3) late 1981. The prototype Tu-134 flew (as the Tu-124A) on 29 July 1963, the production version entering service on 9 September 1967. The Tu-134A entered service in November 1970, followed by the Tu-134B in May 1980.

Notes: The latest series production versions of the Tu-134 are the Tu-134A-3 and B-1 with D-30 Srs III engines offering improved "hot-and-high" performance, the latter subtype have a redesigned and modernised flight deck and spoilers for improved lateral control. The Tu-134A-3 is fitted with lighter seats of new design and has a maximum capacity of 96 passengers. A V-VS Tu-134 is illustrated above.

TUPOLEV TU-134B-1 (CRUSTY)

Dimensions: Span, 95 ft 1¾ in (29,00 m); length, 121 ft 6½ in (37,05 m); height, 30 ft 0 in (9,14 m); wing area, 1,370·3 sq ft (127,30 m²).

TUPOLEV TU-154B-2 (CARELESS)

Country of Origin: USSR.

Type: Medium/long-haul commercial transport.

Power Plant: Three 23,150 lb st (10 500 kgp) Kuznetsov NK-8-2U turbofans.

Performance: Max. cruising speed, 590 mph (950 km/h) at 31,000 ft (9 450 m); econ. cruise, 559 mph (900 km/h) at 36,090 ft (11 000 m); range (with max. payload), 1,710 mls (2 750 km), (with 160 passengers), 2,020 mls (3 250 km), (with 120 passengers), 2,485 mls (4 000 km).

Weights: Max. take-off, 211,644 lb (96 000 kg).

Accommodation: Flight crew of three and basic arrangements for 160 single-class passengers in six-abreast seating, eight first-class and 150 tourist-class passengers, or max. of 169 passengers in high-density configuration.

Status: Prototype Tu-154 flown on 4 October 1968, with first scheduled services operated from 9 February 1972. The initial Tu-154 succeeded in production successively by Tu-154A and Tu-154B, the latter being the principal model at the beginning of 1982, when production was running at approximately four monthly.

Notes: The Tu-154B-2 embodies various changes by comparison with the basic Tu-154B, these including Thomson-CSF/SFIM automatic flight control and navigation equipment, extended wing spoilers, a modest increase in passenger capacity and an additional emergency exit. The Tu-154M (also referred to as the Tu-164), under test in 1981, is a derivative of the Tu-154B-2 from which it differs essentially in having new turbofans of unspecified type which afford a 20% improvement in specific fuel consumption and provide a range (with 154 passengers) of 2,980 miles (4 800 km).

TUPOLEV TU-154B-2 (CARELESS)

Dimensions: Span, 123 ft 2½ in (37,55 m); length, 157 ft 1¾ in (47,90 m); height, 37 ft 4¾ in (11,40 m); wing area, 2,168·92 sq ft (201,45 m²).

YAKOVLEV YAK-36MP (FORGER-A)

Country of Origin: USSR.

Type: Shipboard VTOL air defence and strike fighter.

Power Plant: One (estimated) 17,640 lb st (8 000 kp) lift/cruise turbojet plus two (estimated) 7,935 lb st (3 600 kgp) lift turbojets.

Performance: (Estimated) Max. speed, 695 mph (1 120 km/h) or Mach 1·05 above 36,000 ft (10 970 m), 725 mph (1 167 km/h) or Mach 0·95 at sea level; high-speed cruise, 595 mph (958 km/h) or Mach 0·85 at 20,000 ft (6 095 m); tactical radius (internal fuel and 2,000 lb/900 kg ordnance), 230 mls (370 km) HI-LO-HI, 150 mls (240 km) LO-LO-LO, (reconnaissance mission with recce pod, two drop tanks and two AAMs), 340 mls (547 km).

Weights: (Estimated) Empty, 12,125 lb (5 500 kg); max. take-off, 22,000 lb (9 980 kg).

Armaments: Four underwing pylons with total (estimated) capacity of 2,205 lb (1 000 kg) for bombs, gun or rocket pods, or IR-homing AAMs.

Status: The Yak-36MP is believed to have flown in prototype form in 1971, and to have attained service evaluation status in 1976. At the beginning of 1982, the Yak-36MP was deployed aboard the carriers *Kiev*, *Minsk* and *Novorossiisk*.

Notes: The Yak-36 is unique among current service combat aircraft in that it possesses vertical take-off-and-landing capability but is incapable of performing a rolling take-off. Each of the three Kiev-class carriers includes in its aircraft complement a squadron of 12–15 Yak-36s for fleet defence against shadowing maritime aircraft and anti-shipping strike.

YAKOVLEV YAK-36MP (FORGER-A)

Dimensions: (Estimated) Span, 24 ft 7 in (7,50 m); length, 52 ft 6 in (16,00 m); height, 11 ft 0 in (3,35 m); wing area, 167 sq ft (15,50 m²).

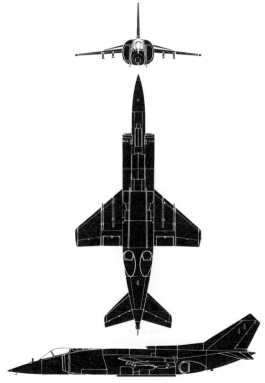

YAKOVLEV YAK-42 (CLOBBER)

Country of Origin: USSR.

Type: Short/medium-haul commercial transport.

Power Plant: Three 14,320 lb st (6 500 kgp) Lotarev D-36 turbofans.

Performance: Max. cruising speed, 503 mph (810 km/h) at 26,245 ft (8 000 m); econ. cruise, 478 mph (770 km/h) at 29,530 ft (9 000 m); long-range cruise, 466 mph (750 km/h); max. cruise altitude, 32,810 ft (10000 m); range (with 31,966 lb/14 500 kg payload), 559 mls (900 km), (with 23,148 lb/10 500 kg payload), 1,243 mls (2 000 km), (with 14,330 lb/6 500 kg payload), 1,864 mls (3 000 km).

Weights: Operational empty, 63,845 lb (28,960 kg); max. take-off, 117,945 lb (53 500 kg).

Accommodation: Basic flight crew of two and various alternative cabin arrangements, including 76 passengers in a mixed-class (16 first class) layout, 100 passengers in a single-class layout with six-abreast seating and 120 passengers in a high-density layout.

Status: First prototype flown on 7 March 1974, with a production prototype following in February 1977. Deliveries to Aeroflot commenced in 1980, and first scheduled service (Moscow–Krasnodar) flown 22 December of that year.

Notes: Intended to operate over relatively short stages and utilise airfields with poor surfaces and limited facilities, the Yak-42 established seven world records in its class for load to altitude in June 1981, and in October 1981 flew a distance of 2,939·7 mls (4 730,9 km) in 7·266 hours, following this on 14–15 December 1981 with a distance of 3,818 mls (6 144,8 km).

YAKOVLEV YAK-42 (CLOBBER)

Dimensions: Span, 114 ft 6 in (34,90 m); length, 119 ft 4 in (36,38 m); height, 32 ft 1⅛ in (9,80 m); wing area, 1,615 sq ft (150,00 m²).

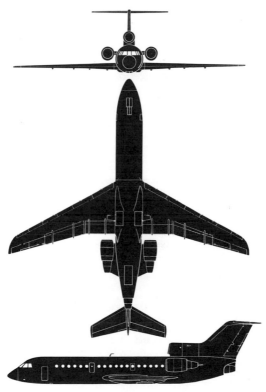

AÉROSPATIALE SA 330J PUMA

Country of Origin: France.
Type: Medium transport helicopter.
Power Plant: Two 1,575 shp Turboméca IVC turboshafts.
Performance: Max. speed, 163 mph (262 km/h); max. continuous cruise at sea level, 159 mph (257 km/h); max. inclined climb, 1,400 ft/min (7,1 m/sec); hovering ceiling (in ground effect), 7,315 ft (2 230 m), (out of ground effect), 4,430 ft (1 350 m); max. range (standard fuel), 342 mls (550 km).
Weights: Empty, 7,969 lb (3 615 kg); max. take-off, 16,534 lb (7 500 kg).
Dimensions: Rotor diam, 49 ft 5¾ in (15,08 m); fuselage length, 46 ft 1½ in (14,06 m).
Notes: The civil SA 330J and the equivalent military SA 330L were the current production models of the Puma at the beginning of 1982 when nearly 750 Pumas of all versions had been ordered. The SA 330J and 330L differ from the civil SA 330F (passenger) and SA 330G (cargo), and SA 330H (military) models that immediately preceded them in having new plastic blades accompanied by increases in gross weights. The SA 330B (French Army), SA 330C (export) and SA 330E (RAF) had 1,328 shp Turmo IIIC4 turboshafts. Components for the Puma are supplied by Westland in the UK (representing approx. 15% of the airframe) and production was five Pumas monthly at the beginning of 1982. The Puma has been delivered to some 45 countries.

AÉROSPATIALE AS 332L SUPER PUMA

Country of Origin: France.
Type: Medium transport helicopter.
Power Plant: Two 1,755 shp Turboméca Makila turboshafts.
Performance: (At 18,080 lb/8 200 kg) Max. speed, 184 mph (296 km/h); max. cruise, 173 mph (278 km/h) at sea level; max. inclined climb, 1,810 ft/min (9,2 m/sec); hovering ceiling (in ground effect), 9,840 ft (3 000 m), (out of ground effect), 7,545 ft (2 300 m); range, 527 mls (850 km).
Weights: Empty, 9,635 lb (4 370 kg); normal loaded, 18,080 lb (8 200 kg); max. take-off, 19,840 lb (9 000 kg).
Dimensions: Rotor diam, 49 ft 5¾ in (15,08 m); fuselage length, 48 ft 7¾ in (14,82 m).
Notes: First flown on 10 October 1980, the AS 332L is a stretched (by 2·5 ft/76 cm) version of the basic Super Puma which is being produced in civil (AS 332C) and military (AS 332B) versions. The AS 332L and M are respectively civil and military variants of the stretched model, and the AS 332F is a navalised ASW version with an overall length of 42 ft 1½ in (12,83 m) with rotor blades folded. Deliveries of the AS 332C began in October 1981 with the AS 332L following in December. Twenty Super Pumas (all versions) delivered by the beginning of 1982, when production was rising to four monthly, this rate being scheduled to increase to five–six by year's end, with 80 to be delivered during course of the year. The AS 332B and C carry 20 troops and 17 passengers respectively.

AÉROSPATIALE SA 342 GAZELLE

Country of Origin: France.

Type: Five-seat light utility helicopter.

Power Plant: One 858 shp Turboméca Astazou XIVH turboshaft.

Performance: Max. speed, 192 mph (310 km/h); max. continuous cruise, 163 mph (263 km/h) at sea level; max. inclined climb, 1,675 ft/min (8,5 m/sec); hovering ceiling (in ground effect), 11,970 ft (3 650 m), (out of ground effect), 9,430 ft (2 875 m); range, 469 mls (754 km) at sea level.

Weights: Empty equipped, 2,149 lb (975 kg); max. take-off (normal), 4,190 lb (1 900 kg).

Dimensions: Rotor diam, 34 ft 5½ in (10,50 m); fuselage length (tail rotor included), 31 ft 2¾ in (9,53 m).

Notes: A more powerful derivative of the SA 341 (592 shp Astazou IIIA), the SA 342 has been exported to Kuwait, Iraq, Libya and other Middle Eastern countries, and may be fitted with four or six HOT missiles, a 20-mm cannon and other weapons for the anti-armour role. One hundred and ten of the French Army's (166) SA 341F Gazelles are being equipped for HOT missiles as SA 341Ms, and 128 SA 342Ms were in process of delivery to the French Army at the beginning of 1982. Versions of the lower-powered SA 341 comprise the SA 341B (British Army), SA 341C (British Navy), SA 341D (RAF), SA 341G (civil) and SA 341H (military export). Sales of SA 341 and 342 Gazelles exceeded 1,000 by the beginning of 1982, when production was continuing in collaboration with Westland.

AÉROSPATIALE AS 350 ECUREUIL

Country of Origin: France.

Type: Six-seat light general-purpose utility helicopter.

Power Plant: (AS 350B) One 641 shp Turboméca Arriel, or (AS 350D) 615 shp Avco Lycoming LTS 101-600A2 turboshaft.

Performance: (AS 350B) Max. speed, 169 mph (272 km/h) at sea level; cruise, 144 mph (232 km/h); max. inclined climb, 1,555 ft/min (7,9 m/sec); hovering ceiling (in ground effect), 9,678 ft (2 950 m), (out of ground effect), 7,382 ft (2 250 m); range, 435 mls (700 km) at sea level.

Weights: Empty, 2,348 lb (1 065 kg); max. take-off, 4,630 lb (2 100 kg).

Dimensions; Rotor diam, 35 ft 0¾ in (10,69 m); fuselage length (tail rotor included), 35 ft 9½ in (10,91 m).

Notes: The first Ecureuil (Squirrel) was flown on 27 June 1974 (with an LTS 101 turboshaft) and the second on 14 February 1975 (with an Arriel). The LTS 101-powered version (AS 350D) is being marked in the USA as the AStar, some 240 having been delivered to US customers by the beginning of 1982, when production rate of both versions was running at 15–20 monthly with more than 500 delivered. The standard Ecureuil is a six-seater and features include composite rotor blades, a so-called Starflex rotor head, simplified dynamic machinery and modular assemblies to simplify changes in the field. The AS 350D AStar version is assembled and finished by Aérospatiale Helicopter at Grand Prairie, Alberta.

AÉROSPATIALE AS 355F ECUREUIL

Country of Origin: France.
Type: Six-seat light general-purpose utility helicopter.
Power Plant: Two 420 shp Allison 250-C20F turboshafts.
Performance: Max. speed, 169 mph (272 km/h) at sea level; max. cruise, 144 mph (232 km/h) at sea level; max. inclined climb, 1,614 ft/min (8,2 m/sec); hovering ceiling (out of ground effect), 7,900 ft (2 410 m); service ceiling, 14,800 ft (4 510 m); range, 470 mls (756 km) at sea level.
Weights: Empty, 2,778 lb (1 260kg); max. take-off, 5,292 lb (2 400 kg).
Dimensions: Rotor diam, 35 ft 0¾ in (10,69 m); fuselage length (tail rotor included), 35 ft 9½ in (10,91 m).
Notes: Flown for the first time on 27 September 1979, the Ecureuil 2 employs an essentially similar airframe and similar dynamic components to those of the single-engined AS 350 Ecureuil (see page 219), and is intended primarily for the North American market on which it is known as the TwinStar with orders totalling 230 by the beginning of 1982. Deliveries of the Ecureuil 2/TwinStar commenced in July 1981, with more than 100 delivered by the beginning of 1982. From the first quarter of 1982, the production model is the AS 355F which possesses a higher maximum take-off weight than the AS 355E that it has succeeded. Total orders for the Ecureuil 2 worldwide exceeded 450 by the beginning of 1982, with production running at 20 monthly.

AÉROSPATIALE SA 361H DAUPHIN

Country of Origin: France.
Type: Light anti-armour helicopter.
Power Plant: One 1,400 shp Turboméca Astazou XXB turbo-shaft.
Performance: Max. cruising speed, 180 mph (289 km/h) at sea level; econ. cruise, 168 mph (270 km/h); max. climb, 2,885 ft/min (14,5 m/sec); hovering ceiling (in ground effect), 12,630 ft (3 850 m); range, 350 mls (565 km).
Weights: Max. take-off, 7,496 lb (3 400 kg).
Dimensions: Rotor diam, 38 ft 4 in (11,68 m); fuselage length, 40 ft 3 in (12,27 m).
Notes: The SA 361H/HCL (*hélicoptère de combat léger*) is an anti-armour version of the basic SA 361H military variant of the Dauphin (its civil equivalent being the SA 361F). It is equipped with a forward-looking infra-red aiming system and up to eight HOT (High-subsonic Optically-guided Tube-launched) anti-armour missiles, but retains its capability to transport up to 13 fully-equipped troops. The first prototype Dauphin flew on 2 June 1972, production being initiated as the SA 360, and the prototype of the SA 361, an overpowered version intended specifically for hot-and-high operating conditions, followed on 12 July 1976. If ordered, the HCL model can be delivered in 1982. The SA 360 Dauphin is powered by a 1,050 shp Astazou XVIIIA and normally carries a pilot and nine passengers. An alternative layout provides accommodation for 14.

221

AÉROSPATIALE SA 365 DAUPHIN 2

Country of Origin: France.
Type: Multi-purpose and transport helicopter.
Power Plant: Two 700 shp Turboméca Arriel 1C turboshafts.
Performance: (SA 365N) Max. speed, 190 mph (306 km/h); max. continuous cruise, 173 mph (278 km/h) at sea level; max. inclined climb, 1,279 ft/min (6,5 m/sec); hovering ceiling (in ground effect), 3,296 ft (1 005 m), (out of ground effect), 3,116 ft (950 m); range, 548 mls (882 km) at sea level.
Weights: Empty, 4,288 lb (1 945 kg); max. take-off, 8,487 lb (3 850 kg).
Dimensions: Rotor diam, 39 ft 1½ in (11,93 m); fuselage length (including tail rotor), 37 ft 6⅓ in (11,44 m).
Notes: Flown as a prototype on 31 March 1979, the SA 365 is the latest derivative of the basic Dauphin, and is being manufactured in four versions, the 10–14 seat commercial SA 365N, the military SA 365M which can transport 13 commandos and carry eight HOT missiles, the navalised SA 365F with folding rotor, Agrion radar and four AS 15TT anti-ship missiles (20 ordered by Saudi Arabia for delivery from 1983) and the SA 366G, an Avco Lycoming LTS 101-750-powered search and rescue version for the US Coast Guard (illustrated above) as the HH-65A Seaguard. Ninety of the last version are to be procured by the US Coast Guard from 1982, with completion in 1985. Production of the SA 365N is scheduled to attain 10 monthly by mid-1982 and rise to 16 monthly by the beginning of 1983.

AGUSTA A 109A Mᴋ II

Country of Origin: Italy.
Type: Eight-seat light utility helicopter.
Power Plant: Two 420 shp Allison 250-C20B turboshafts.
Performance: (At 5,402 lb/2 450 kg) Max. speed, 193 mph
(311 km/h); max. continuous cruise, 173 mph (278 km/h);
range cruise, 143 mph (231 km/h); max. inclined climb rate,
1,820 ft/min (9,25 m/sec); hovering ceiling (in ground effect),
9,800 ft (2 987 m), (out of ground effect), 6,800 ft (2 073 m);
max. range, 356 mls (573 km).
Weights: Empty equipped, 3,125 lb (1 418 kg); max. take-off,
5,730 lb (2 600 kg).
Dimensions: Rotor diam, 36 ft 1 in (11,00 m); fuselage length,
35 ft 2½ in (10,73 m).
Notes: The A 109A Mk II is an improved model of the basic A
109A, the first of four prototypes of which flew on 4 August
1971, with customer deliveries commencing late 1976. Some
300 A 109AS had been ordered by the beginning of 1982, at
which time more than 200 had been delivered with production
running at six–seven monthly. The Mk II, which supplanted the
initial model in production during 1981, has been the subject of
numerous detail improvements, the transmission rating of the
combined engines being increased from 692 to 740 shp, and the
maximum continuous rating of each engine from 385 to 420 shp.
A multi-role military version has been procured by Italian, Argentine and Libyan forces.

223

BELL MODEL 206B JETRANGER III

Country of Origin: USA.
Type: Five-seat light utility helicopter.
Power Plant: One 420 shp Allison 250-C20B turboshaft.
Performance: (At 3,200 lb/1 451 kg) Max. speed, 140 mph (225 km/h) at sea level; max. cruise, 133 mph (214 km/h) at sea level; hovering ceiling (in ground effect), 12,700 ft (3 871 m), (out of ground effect), 6,000 ft (1 829 m); max. range (no reserve), 360 mls (579 km).
Weights: Empty, 1,500 lb (680 kg); max. take-off, 3,200 lb (1 451 kg).
Dimensions: Rotor diam, 33 ft 4 in (10,16 m); fuselage length, 31 ft 2 in (9,50 m).
Notes: Introduced in 1977, with deliveries commencing in July of that year, the JetRanger III differs from the JetRanger II which it supplants in having an uprated engine, an enlarged and improved tail rotor mast and more minor changes. Some 2,700 commercial JetRangers had been delivered by the beginning of 1982, both commercial and military versions (including production by licensees) totalling more than 6,500. A light observation version of the JetRanger for the US Army is designated OH-58 Kiowa and a training version for the US Navy is known as the TH-57A SeaRanger. The JetRanger is built by Agusta in Italy as the AB 206, and at the beginning of 1982, Agusta was producing the JetRanger at a rate of six monthly with approximately 1,000 delivered.

BELL MODEL 206L-1 LONGRANGER II

Country of Origin: USA.

Type: Seven-seat light utility helicopter.

Power Plant: One 500 shp Allison 250-C28B turboshaft.

Performance: (At 3,900 lb/1 769 kg) Max. speed, 144 mph (232 km/h); cruise, 136 mph (229 km/h) at sea level; hovering ceiling (in ground effect), 8,200 ft (2 499 m), (out of ground effect), 2,000 ft (610 m); range, 390 mls (628 km) at sea level, 430 mls (692 km) at 5,000 ft (1 524 m).

Weights: Empty, 2,160 lb (980 kg); max. take-off, 4,150 lb (1 882 kg).

Dimensions: Rotor diam, 37 ft 0 in (11,28 m); fuselage length, 33 ft 3 in (10,13 m).

Notes: The Model 206L-1 LongRanger II is a stretched and more powerful version of the Model 206B JetRanger III, with longer fuselage, increased fuel capacity, an uprated engine and a larger rotor. The LongRanger is being manufactured in parallel with the JetRanger III and initial customer deliveries commenced in October 1975, prototype testing having been initiated on 11 September 1974. The LongRanger is available with emergency flotation gear and with a 2,000-lb (907-kg) capacity cargo hook. In the aeromedical or rescue role the LongRanger can accommodate two casualty stretchers and two ambulatory casualties. The 206L-1 LongRanger II was introduced in 1978, and production was 15 monthly at the beginning of 1982, with more than 700 LongRangers delivered.

BELL MODEL 206L TEXASRANGER

Country of Origin: USA.

Type: Multi-role military helicopter.

Power Plant: One 650 shp Allison 250-C30P turboshaft.

Performance: (Utility configuration) Max. cruising speed, 131 mph (211 km/h); econ. cruise, 129 mph (207 km/h); max. inclined climb, 1,360 ft/min (6,9 m/sec); hovering ceiling (in ground effect), 12,000 ft (3 658 m); max. range (standard fuel), 356 mls (573 km).

Weights: Max. take-off (internal load), 4,150 lb (1 882 kg), (external jettisonable load), 4,250 lb (1 928 kg).

Dimensions: Rotor diam, 37 ft 0 in (11,28 m); fuselage length, 33 ft 3 in (10,13 m).

Notes: The TexasRanger is a multi-role military version of the Model 206L-1 LongRanger II (see page 225) introduced in 1980. The TexasRanger can carry four TOW (Tube-launched Optically-tracked Wire-guided) missiles for the anti-armour role, or two twin 7,62-mm gun pods with 500 rpg, or two pods each containing seven 2·75-in (70-mm) rockets. Armoured crew seats are fitted and missile control electronics have a special, readily removable and reinstatable rear-seat pallet mounting, permitting conversion for utility role within 15 minutes. With the weapons pallet removed, the TexasRanger can carry seven persons, including the crew, and the helicopter is configured for quick-change to cover a variety of roles, including troop lift, command control, armed reconnaissance and surveillance.

BELL AH-1S HUEYCOBRA

Country of Origin: USA.

Type: Two-seat light attack helicopter.

Power Plant: One 1,800 shp Avco Lycoming T53-L-703 turboshaft.

Performance: Max. speed, 172 mph (277 km/h), (TOW configuration), 141 mph (227 km/h); max. inclined climb, 1,620 ft/min (8,23 m/sec); hovering ceiling TOW configuration (in ground effect), 12,200 ft (3 720 m); max. range, 357 mls (574 km).

Weights: (TOW configuration) Operational empty, 6,479 lb (2 939 kg); max. take-off, 10,000 lb (4 535 kg).

Dimensions: Rotor diam, 44 ft 0 in (13,41 m); fuselage length, 44 ft 7 in (13,59 m).

Notes: The AH-1S is a dedicated attack and anti-armour helicopter serving primarily with the US Army which had received 297 new-production AH-1S HueyCobras by mid-1981, plus 290 resulting from the conversion of earlier AH-1G and AH-1Q HueyCobras. Current planning calls for conversion of a further 372 AH-1Gs to AH-1S standards, and both conversion and new-production AH-1S HueyCobras are being progressively upgraded to "Modernised AH-1S" standard, the entire programme being scheduled for completion in 1985, resulting in a total of 959 "Modernised" AH-1S HueyCobras. In December 1979, one YAH-1S was flown with a four-bladed main rotor as the Model 249.

BELL AH-1T SEACOBRA

Country of Origin: USA.
Type: Two-seat light attack helicopter.
Power Plant: One 1,970 shp Pratt & Whitney T400-WV-402 coupled turboshaft.
Performance: (Attack configuration at 12,401 lb/5 625 kg) Max. speed, 181 mph (291 km/h) at sea level; average cruise, 168 mph (270 km/h); max. inclined climb, 2,190 ft/min (11,12 m/sec); hovering ceiling (out of ground effect), 5,350 ft (1 630 m); range, 276 mls (445 km).
Weights: Empty, 8,030 lb (3 642 kg); max. take-off, 14,000 lb (6 350 kg).
Dimensions: Rotor diam, 48 ft 0 in (14,63 m); fuselage length, 45 ft 3 in (13,79 m).
Notes: The SeaCobra is a twin-turboshaft version of the Huey-Cobra (see page 227), the initial model for the US Marine Corps having been the AH-1J (69 delivered of which two modified as AH-1T prototypes). The AH-1T features uprated components for significantly increased payload and performance, the first example having been delivered to the US Marine Corps on 15 October 1977, and a further 56 being delivered to the service of which 23 being modified to TOW configuration. The AH-1T has a three-barrel 20-mm cannon barbette under the nose, and four stores stations under the stub wings for seven- or 19-tube launchers, Minigun pods, etc.

228

BELL MODEL 214ST

Country of Origin: USA.

Type: Medium transport helicopter (19 seats).

Power Plant: Two 1,625 shp (limited to combined output of 2,250 shp) General Electric CT7-2 turboshafts.

Performance: Max. crusing speed, 164 mph (264 km/h) at sea level, 161 mph (259 km/h) at 4,000 ft (1 220 m); hovering ceiling (in ground effect), 12,600 ft (3 840 m), (out of ground effect), 3,300 ft (1 005 m); range (standard fuel), 460 mls (740 km).

Weights: Max. take-off (internal load), 15,500 lb (7 030 kg), (external jettisonable load), 16,500 lb (7 484 kg).

Dimensions: Rotor diam, 52 ft 0 in (15,85 m); fuselage length, 50 ft 0 in (15,24 m).

Notes: The Model 214ST (Super Transport) is a significantly improved derivative of the Model 214B BigLifter (see 1978 edition), production of which was phased out early 1981, and initial customer deliveries are scheduled for early 1982. The Model 214ST test-bed was first flown in March 1977, and the first of three representative prototypes (one in military configuration and two for commercial certification) commenced its test programme in August 1979. Work on an initial series of 100 helicopters of this type commenced in 1981, with production tempo scheduled to reach three monthly in 1982. Alternative layouts are available for either 16 or 17 passengers.

BELL MODEL 222

Country of Origin: USA.
Type: Eight/ten-seat light utility and transport helicopter.
Power Plant: Two 620 shp Avco Lycoming LTS 101-650C-2 turboshafts.
Performance: Max. cruising speed, 150 mph (241 km/h) at sea level, 146 mph (235 km/h) at 8,000 ft (2 400 m); max. climb, 1,730 ft/min (8,8 m/sec); hovering ceiling (in ground effect), 10,300 ft (3 135 m), (out of ground effect), 6,400 ft (1 940 m); range (no reserves), 450 mls (724 km) at 8,000 ft (2 400 m).
Weights: Empty equipped, 4,577 lb (2 076 kg); max. take-off (standard configuration), 7,650 lb (3 470 kg).
Dimensions: Rotor diam, 39 ft 9 in (12,12 m); fuselage length, 39 ft 9 in (12,12 m).
Notes: The first of five prototypes of the Model 222 was flown on 13 August 1976, an initial production series of 250 helicopters of this type being initiated in 1978, with production deliveries commencing in January 1980, and some 200 delivered by beginning of 1982, when orders totalled approximately 250. Several versions of the Model 222 are on offer or under development, these including an executive version with a flight crew of two and five or six passengers and the so-called "offshore" model with accommodation for eight passengers and a flight crew of two. Options include interchangeable skids. The Model 222 is claimed to have been the first twin-engined light commercial helicopter built in the USA.

BELL MODEL 412

Country of Origin: USA.
Type: Fifteen-seat utility transport helicopter.
Power Plant: One 1,800 shp (1,308 shp take-off rating) Pratt & Whitney PT6T-3B turboshaft.
Performance: Max. speed, 149 mph (240 km/h) at sea level; cruise, 143 mph (230 km/h) at sea level, 146 mph (235 km/h) at 5,000 ft (1 525 m); hovering ceiling (in ground effect), 10,800 ft (3 290 m), (out of ground effect), 7,100 ft (2 165 m) at 10,500 lb/4 763 kg; max. range, 282 mls (454 km), (with auxiliary tanks), 518 mls (834 km).
Weights: Empty equipped, 6,070 lb (2 753 kg); max. take-off, 11,500 lb (5 216 kg).
Dimensions: Rotor diam, 46 ft 0 in (14,02 m); fuselage length, 41 ft 8½ in (12,70 m).
Notes: The Model 412, flown for the first time in August 1979, is an updated Model 212 (production of which was completed early 1981) with a new-design four-bladed rotor, a shorter rotor mast assembly, and uprated engine and transmission systems, giving more than twice the life of the Model 212 units. Composite rotor blades are used and the rotor head incorporates elastomeric bearings and dampers to simplify moving parts. A third prototype joined the test programme in 1980, an initial series of 200 helicopters being laid down with initial deliveries commencing February 1981, and licence manufacture being undertaken by Agusta in Italy, with initial deliveries early 1982.

BOEING VERTOL 234 CHINOOK

Country of Origin: USA.

Type: Commercial transport helicopter.

Power Plant: Two 4,075 shp Avco Lycoming AL 5512 turboshafts.

Performance: Max. cruising speed (at 47,000 lb/21 318 kg), 167 mph (269 km/h) at 2,000 ft (610 m); range cruise, 155 mph (250 km/h); max. inclined climb, 1,350 ft/min (6,8 m/sec); hovering ceiling (in ground effect), 9,150 ft (2 790 m), (out of ground effect), 4,900 ft (1 495 m); range (44 passengers and 45 min reserves), 627 mls (1 010 km), (max. fuel), 852 mls (1 371 km).

Weights: Empty, 24,449 lb (11 090 kg); max. take-off, 47,000 lb (21 318 kg).

Dimensions: Rotor diam (each), 60 ft 0 in (18,29 m); fuselage length, 52 ft 1 in (15,87 m).

Notes: Possessing an airframe based on the latest Model 414 military Chinook (see opposite), the Model 234 has been developed specifically for commercial purposes and two basic versions are offered, a long-range model described above and a utility model with fuel tank-housing side fairings removed. The first Model 234 was flown on 19 August 1980, certification being obtained mid-1981 and the first deliveries (to British Airways Helicopters with six on order) followed during the course of the year, primarily for North Sea oil rig support.

BOEING VERTOL 414 CHINOOK HC Mk 1

Country of Origin: USA.
Type: Medium transport helicopter.
Power Plant: Two 3,750 shp Avco Lycoming T55-L-11E turboshafts.
Performance: (At 45,400 lb/20 593 kg) Max. speed, 146 mph (235 km/h) at sea level; average cruise, 131 mph (211 km/h); max. inclined climb, 1,380 ft/min (7,0 m/sec); service ceiling, 8,400 ft (2 560 m); max. ferry range, 1,190 mls (1 915 km).
Weights: Empty, 22,591 lb (10 247 kg); max. take-off, 50,000 lb (22 680 kg).
Dimensions: Rotor diam (each), 60 ft 0 in (18,29 m); fuselage length, 51 ft 0 in (15,55 m).
Notes: The Model 414 as supplied to the RAF as the Chinook HC Mk 1 combines some features of the US Army's CH-47D (see 1980 edition) and features of the Canadian CH-147, but with provision for glassfibre/carbonfibre rotor blades. The first of 33 Chinook HC Mk 1s for the RAF was flown on 23 March 1980 and accepted on 2 December 1980, with deliveries continuing through 1981. The RAF version can accommodate 44 troops and has three external cargo hooks. Boeing Vertol manufactured 32 Chinooks (mostly HC Mk 1s for the RAF) during 1981, and initiated the conversion to essentially similar CH-47D standards a total of 436 CH-47As, Bs and Cs. Licence manufacture of the Chinook is undertaken in Italy.

HUGHES 500MD DEFENDER II

Country of Origin: USA.
Type: Light gunship and multi-role helicopter.
Power Plant: One 420 shp Allison 250-C20B turboshaft.
Performance: (At 3,000 lb/1 362 kg) Max. speed, 175 mph (282 km/h) at sea level; cruise, 160 mph (257 km/h) at 4,000 ft (1 220 m); max. inclined climb, 1,920 ft/min (9,75 m/sec); hovering ceiling (in ground effect), 8,800 ft (2 682 m), (out of ground effect), 7,100 ft (2 164 m); max. range, 263 mls (423 km).
Weights: Empty, 1,295 lb (588 kg); max. take-off (internal load), 3,000 lb (1 362 kg), (with external load), 3,620 lb (1 642 kg).
Dimensions: Rotor diam, 26 ft 5 in (8,05 m); fuselage length, 21 ft 5 in (6,52 m).
Notes: The Defender II multi-mission version of the Model 500MD was introduced mid-1980 for 1982 delivery, and features a Martin Marietta rotor mast-top sight, a General Dynamics twin-Stinger air-to-air missile pod, an underfuselage 30-mm chain gun and a pilot's night vision sensor. The Defender II can be rapidly reconfigured for anti-armour target designation, anti-helicopter, suppressive fire and transport roles. The Model 500MD TOW Defender (carrying four tube-launched optically-tracked wire-guided anti-armour missiles) is currently in service with Israel (30), South Korea (25) and Kenya (15). Production of the 500 was 17–18 monthly at beginning of 1982.

HUGHES AH-64 APACHE

Country of Origin: USA.

Type: Tandem two-seat attack helicopter.

Power Plant: Two 1,536 shp General Electric T700-GE-700 turboshafts.

Performance: Max. speed, 191 mph (307 km/h); cruise, 179 mph (288 km/h); max. inclined climb, 3,200 ft/min (16,27 m/sec); hovering ceiling (in ground effect), 14,600 ft (4 453 m), (outside ground effect), 11,800 ft (3 600 m); service ceiling, 8,000 ft (2 400 m); max. range, 424 mls (682 km).

Weights: Empty, 9,900 lb (4 490 kg); primary mission, 13,600 lb (6 169 kg); max. take-off, 17,400 lb (7 892 kg).

Dimensions: Rotor diam, 48 ft 0 in (14,63 m); fuselage length, 49 ft 4½ in (15,05 m).

Notes: Winning contender in the US Army's AAH (Advanced Attack Helicopter) contest, the YAH-64 flew for the first time on 30 September 1975. Two prototypes were used for the initial trials, the first of three more with fully integrated weapons systems commenced trials on 31 October 1979, a further three following in 1980. Planned total procurement comprises 446 AH-64s. The AH-64 is armed with a single-barrel 30-mm gun based on the chain-driven bolt system and suspended beneath the forward fuselage, and eight BGM-71A TOW anti-armour missiles may be carried, alternative armament including 16 Hellfire laser-seeking missiles. Target acquisition and designation and a pilot's night vision system will be used.

KAMOV KA-25 (HORMONE A)

Country of Origin: USSR.

Type: Shipboard anti-submarine warfare helicopter.

Power Plant: Two 900 shp Glushenkov GTD-3 turboshafts.

Performance: (Estimated) Max. speed, 130 mph (209 km/h); normal cruise, 120 mph (193 km/h); max. range, 400 mls (644 km); service ceiling, 11,000 ft (3 353 m).

Weights: (Estimated) Empty, 10,500 lb (4 765 kg); max. take-off, 16,500 lb (7 484 kg).

Dimensions: Rotor diam (each), 51 ft 7½ in (15,74 m); approx. fuselage length, 35 ft 6 in (10,82 m).

Notes: Possessing a basically similar airframe to that of the Ka-25K (see 1973 edition) and employing a similar self-contained assembly comprising rotors, transmission, engines and auxiliaries, the Ka-25 serves with the Soviet Navy primarily in the ASW role but is also employed in the utility and transport roles. The ASW Ka-25 serves aboard the helicopter cruisers *Moskva* and *Leningrad*, and the carriers *Kiev* and *Minsk*, as well as with shore-based units. A search radar installation is mounted in a nose randome, but other sensor housings and antennae differ widely from helicopter to helicopter. There is no evidence that externally-mounted weapons may be carried. Each landing wheel is surrounded by an inflatable pontoon surmounted by inflation bottles. The Hormone-A is intended for ASW operations where the Hormone-B is used for over-the-horizon missile targeting.

MBB BO 105L

Country of Origin: Federal Germany.
Type: Five/six-seat light utility helicopter.
Power Plant: Two 550 shp Allison 250-C28C turboshafts.
Performance: Max. speed, 168 mph (270 km/h) at sea level; max. cruise, 157 mph (252 km/h) at sea level; max. climb, 1,970 ft/min (10 m/sec); hovering ceiling (in ground effect), 13,120 ft (4 000 m), (out of ground effect), 11,280 ft (3 440 m); range, 286 mls (460 km).
Weights: Empty, 2,756 lb (1 250 kg); max. take-off, 5,291 lb (2 400 kg), (with external load), 5,512 lb (2 500 kg).
Dimensions: Rotor diam, 32 ft 3½ in (9,84 m); fuselage length, 28 ft 1 in (8,56 m).
Notes: The BO 105L is a derivative of the BO 105CB (see 1979 edition) with uprated transmission and more powerful turboshaft for "hot-and-high" conditions. It is otherwise similar to the BO 105CB (420 shp Allison 250-C20B) which was continuing in production at the beginning of 1982, when more than 600 BO 105s (all versions) had been delivered and production was running at 10–12 monthly, and licence assembly was being undertaken in Indonesia, the Philippines and Spain. Deliveries to the Federal German Army of 227 BO 105M helicopters for liaison and observation tasks commenced late 1979, and deliveries of 212 HOT-equipped BO 105s (illustrated) for the anti-armour role began on 4 December 1980. The latter have uprated engines and transmission systems.

MBB-KAWASAKI BK 117

Countries of Origin: Federal Germany and Japan.
Type: Multi-purpose eight-to-twelve-seat helicopter.
Power Plant: Two 600 shp Avco Lycoming LTS 101-650B-1
turboshafts.
Performance: Max. speed, 171 mph (275 km/h) at sea level;
cruise, 164 mph (264 km/h) at sea level: max. climb, 1,970 ft/
min (10 m/sec); hovering ceiling (in ground effect), 13,450 ft
(4 100 m), (out of ground effect), 10,340 ft (3 150 m); range
(max. payload), 339 mls (545,4 km).
Weights: Empty, 3,351 lb (1 520 kg); max. take-off, 6,173 lb
(2 800 kg).
Dimensions: Rotor diam, 36 ft 1 in (11,00 m); fuselage length,
32 ft 5 in (9,88 m).
Notes: The BK 117 is a co-operative development between
Messerschmitt-Bölkow-Blohm and Kawasaki, the first of two
flying prototypes commencing its flight test programme on 13
June 1979 (in Germany), with the second following on 10
August (in Japan). A decision to proceed with series production
was taken in 1980, with production deliveries commencing first
quarter of 1982. Some 150 BK 117s had been ordered by the
beginning of 1982. MBB is responsible for the main and tail rotor
systems, tail unit and hydraulic components, while Kawasaki is
responsible for production of the fuselage, undercarriage, transmission and some other components. Several military versions
are currently proposed.

MIL MI-8 (HIP)

Country of Origin: USSR.
Type: Assault transport helicopter.
Power Plant: Two 1,700 shp Isotov TV2-117A turboshafts.
Performance: Max. speed, 161 mph (260 km/h) at 3,280 ft (1 000 m), 155 mph (250 km/h) at sea level; max. cruise, 140 mph (225 km/h); hovering ceiling (in ground effect), 6,233 ft (1 900 m), (out of ground effect), 2,625 ft (800 m); range (standard fuel), 290 mls (465 km).
Weights: (Hip-C) Empty, 14,603 lb (6 624 kg); normal loaded, 24,470 lb (11 100 kg); max. take-off, 26,455 lb (12 000 kg).
Dimensions: Rotor diam, 69 ft 10$\frac{1}{4}$ in (21,29 m); fuselage length, 60 ft 0$\frac{3}{4}$ in (18,31 m).
Notes: Currently being manufactured at a rate of 700–800 annually, with 6,000–7,000 delivered for civil and military use since its debut in 1961, the Mi-8 is numerically the most important Soviet helicopter. Current military versions include the Hip-C basic assault transport, the Hip-D with additional antennae and podded equipment for electronic tasks, the Hip-E and the Hip-F, the former carrying up to six rocket pods and four Swatter IR-homing anti-armour missiles, and the latter carrying six Sagger wire-guided anti-armour missiles. The Mi-8 can accommodate 24 troops or 12 stretchers, and most have a 12,7-mm machine gun in the nose. Commercial models include the basic 28–32 passenger model and the Mi-8T utility version. An enhanced version, the Mi-17, is described on page 241.

MIL MI-14 (HAZE-A)

Country of Origin: USSR.

Type: Amphibious anti-submarine helicopter.

Power Plant: Two 1,500 shp Isotov TV-2 turboshafts.

Performance: (Estimated) Max. speed, 143 mph (230 km/h); max. cruise, 130 mph (210 km/h); hovering ceiling (in ground effect), 5,250 ft (1 600 m), (out of ground effect), 2,295 ft (700 m); tactical radius, 124 mls (200 km).

Weights: (Estimated) Max. take-off, 26,455 lb (12 000 kg).

Dimensions: Rotor diam, 69 ft 10¼ in (21,29 m); fuselage length, 59 ft 7 in (18,15 m).

Notes: The Mi-14 amphibious anti-submarine warfare helicopter, which serves with shore-based elements of the Soviet Naval Air Force, is a derivative of the Mi-8 (see page 239) with essentially similar power plant and dynamic components, and much of the structure is common between the two helicopters. New features include the boat-type hull, outriggers which, housing the retractable lateral twin-wheel undercarriage members, incorporate water rudders, a search radar installation beneath the nose and a sonar "bird" beneath the tailboom root. The Mi-14 may presumably be used for over-the-horizon missile targeting and for such tasks as search and rescue. It may also be assumed that the Mi-14 possesses a weapons bay for ASW torpedoes, nuclear depth charges and other stores. This amphibious helicopter reportedly entered service in 1975 and about 100 were in Soviet Navy service by the beginning of 1982.

MIL MI-17

Country of Origin: USSR.
Type: Medium transport helicopter.
Power Plant: Two 1,900 shp Isotov TV3-117MT turboshafts.
Performance: (At 28,660 lb/13 000 kg) Max. speed, 162 mph (260 km/h); max. continuous cruise, 149 mph (240 km/h) at sea level; hovering ceiling (at 24,250 lb/11 000 kg out of ground effect), 5,800 ft (1 770 m); max. range, 777 mls (1 250 km).
Weights: Empty, 15,652 lb (7 100 kg); normal loaded, 24,250 lb (11 000 kg); max. take-off, 28,660 lb (13 000 kg).
Dimensions: Rotor diam, 69 ft 10¼ in (21,29 m); fuselage length, 60 ft 5¼ in (18,42 m).
Notes: The Mi-17 medium-lift helicopter is essentially a more powerful and modernised derivative of the late 'fifties technology Mi-8 (see page 239). The airframe and rotor are essentially unchanged, apart from some structural reinforcement of the former, but higher-performance turboshafts afford double the normal climb rate and out-of-ground-effect hover ceiling of the earlier helicopter, and increase permissible maximum take-off weight. The Mi-17 has a crew of two–three and can accommodate 24 passengers, 12 casualty stretchers or up to 8,818 lb (4 000 kg) of freight. The TV3-117 turboshafts utilised by the Mi-17 are also believed to be installed in late production version of the military Mi-8 (eg, the Hip-E and Hip-F). Externally, the Mi-17 is virtually indistinguishable from its precursor, the Mi-8, apart from marginally shorter engine nacelles.

MIL MI-24 (HIND-D)

Country of Origin: USSR.

Type: Assault and anti-armour helicopter.

Power Plant: Two 2,200 shp Isotov TV3-117 turboshafts.

Performance: (Estimated) Max. speed, 170–180 mph (273–290 km/h) at 3,280 ft (1 000 m); max. cruise, 145 mph (233 km/h); max. inclined climb rate, 3,000 ft/min (15,24 m/sec).

Weights: (Estimated) Normal take-off, 22,000 lb (10 000 kg).

Dimensions: (Estimated) Rotor diam, 55 ft 0 in (16,76 m); fuselage length, 55 ft 6 in (16,90 m).

Notes: By comparison with the Hind-A version of the Mi-24 (see 1977 edition), the Hind-D embodies a redesigned forward fuselage and is optimised for the gunship role, having tandem stations for the weapons operator (in nose) and pilot. The Hind-D can accommodate eight fully-equipped troops, has a barbette-mounted four-barrel rotary-type 12,7-mm cannon beneath the nose and can carry up to 2,800 lb (1 275 kg) of ordnance externally, including four AT-2 Swatter IR-homing anti-armour missiles and four pods each with 32 57-mm rockets. It has been exported to Afghanistan, Algeria, Bulgaria, Czechoslovakia, East Germany, Hungary, Iraq, Libya, Poland and South Yemen. The Hind-E is similar but has provision for four laser-homing tube-launched Spiral anti-armour missiles, and embodies some structural hardening, steel and titanium being substituted for aluminium in certain critical components. More than 500 Hind-D and -E deployed by WarPac.

MIL MI-26 (HALO)

Country of Origin: USSR.

Type: Military and commercial heavy-lift helicopter.

Power Plant: Two 11,400 shp Lotarev D-136 turboshafts.

Performance: Max. speed, 183 mph (295 km/h); normal cruise, 158 mph (255 km/h); hovering ceiling (in ground effect), 14,765 ft (4 500 m), (out of ground effect), 5,905 ft (1 800 m); range (at 109,127 lb/49 500 kg), 310 mls (500 km), (at 123,457 lb/56 000 kg), 497 mls (800 km).

Weights: Empty, 62,169 lb (28 200 kg); normal load, 109,227 lb (49 500 kg); max. take-off, 123,457 lb (56 000 kg).

Dimensions: Rotor diam, 104 ft $11\frac{7}{8}$ in (32,00 m); fuselage length (nose to tail rotor), 110 ft $7\frac{3}{4}$ in (33,73 m).

Notes: The heaviest and most powerful helicopter ever flown, the Mi-26 first flew as a prototype on 14 December 1977, production of pre-series machines commencing in 1980, and preparations for full-scale production having begun in 1981. Featuring an innovative eight-bladed main rotor and carrying a flight crew of five, the Mi-26 has a max. internal payload of 44,090 lb (20 000 kg). The freight hold is larger than that of the fixed-wing Antonov An-12 transport and at least 70 combat-equipped troops or 40 casualty stretchers can be accommodated. Although allegedly developed to a civil requirement, the primary role of the Mi-26 is obviously military and it is anticipated that the Soviet Air Force will achieve initial operational capability with the series version in 1984–85.

ROBINSON R22 HP

Country of Origin: USA.
Type: Two-seat light utility helicopter.
Power Plant: One 124 hp (derated from 160 hp) Avco Lycoming O-320-B2G four-cylinder horizontally-opposed engine.
Performance: Max. speed, 116 mph (187 km/h); cruise (65% power), 108 mph (174 km/h); max. inclined climb, 1,200 ft/min (6,0 m/sec); hovering ceiling (in ground effect), 8,300 ft (2 530 m), (out of ground effect), 6 400 ft (1 950 m); max. range, 240 mls (386 km).
Weights: Empty, 790 lb (358 kg); max. take-off, 1,300 lb (590 kg).
Dimensions: Rotor diam, 25 ft 2 in (7,67 m); fuselage length, 20 ft 8 in (6,30 m).
Notes: The prototype of the R22 flew for the first time on 28 August 1975, deliveries of production helicopters commencing in October 1979, and some 300 had been delivered by the beginning of 1982, when production was averaging one per day. With the two hundredth production helicopter, a higher compression engine was introduced as standard. With the modified engine, the helicopter is known as the R22 HP and offers a marginally higher cruising speed and appreciably enhanced hovering ceiling, both in and out of ground effect. The R22 is claimed to be one of the least expensive helicopters currently in production, and orders for some 600 had been placed by the beginning of 1982.

244

SIKORSKY S-61D (SEA KING)

Country of Origin: USA.
Type: Amphibious anti-submarine helicopter.
Power Plant: Two 1,500 shp General Electric T58-GE-10 turboshafts.
Performance: Max. speed, 172 mph (277 km/h) at sea level; inclined climb, 2,200 ft/min (11,2 m/sec); hovering ceiling (out of ground effect), 8,200 ft (2 500 m); range (with 10% reserves), 622 mls (1 000 km).
Weights: Empty equipped, 12,087 lb (5 481 kg); max. take-off, 20,500 lb (9 297 kg).
Dimensions: Rotor diam, 62 ft 0 in (18,90 m); fuselage length, 54 ft 9 in (16,69 m).
Notes: A more powerful derivative of the S-61B, the S-61D serves with the US Navy, as the SH-3D, 72 helicopters of this type following on production of 255 SH-3As (S-61Bs) for the ASW role for the US Navy, four being supplied to the Brazilian Navy and 22 to the Spanish Navy. Four similar aircraft have been supplied to the Argentine Navy as S-61D-4s and 11 have been supplied to the US Army/US Marine Corps Executive Flight Detachment as VH-3Ds. Licence manufacture of the S-61D is being undertaken in the United Kingdom (see page 250), in Japan for the Maritime Self-Defence Force and in Italy by Agusta for the Italian Navy and for export. The SH-3G and SH-3H are upgraded versions of the SH-3A. Manufacture by the parent company has now ceased.

SIKORSKY CH-53E SUPER STALLION

Country of Origin: USA.

Type: Amphibious assault transport helicopter.

Power Plant: Three 4,380 shp General Electric T64-GE-415 turboshafts.

Performance: (At 56,000 lb/25 400 kg) Max. speed, 196 mph (315 km/h) at sea level; cruise, 173 mph (278 km/h) at sea level; max. inclined climb, 2,750 ft/min (13,97 m/sec); hovering ceiling (in ground effect), 11,550 ft (3 520 m), (out of ground effect), 9,500 ft (2 895 m); range, 1,290 mls (2 075 km).

Weights: Empty, 32,878 lb (14 913 kg); max. take-off, 73,500 lb (33 339 kg).

Dimensions: Rotor diam, 79 ft 0 in (24,08 m); fuselage length, 73 ft 5 in (22,38 m).

Notes: The CH-53E is a growth version of the CH-53D Sea Stallion (see 1974 edition) embodying a third engine, an uprated transmission system, a seventh main rotor blade and increased rotor diameter. The first of two prototypes was flown on 1 March 1974, and the first of two pre-production examples followed on 8 December 1975, successive production orders totalling 49 helicopters to be divided between the US Navy (16) and US Marine Corps (33), the first production Super Stallion having made its first flight on 13 December 1980. The CH-53E can accommodate up to 55 troops in a high-density seating arrangement. Fleet deliveries began mid-1981, and 15 had been accepted by the USN and USMC by beginning of 1982.

SIKORSKY S-70 (UH-60A) BLACK HAWK

Country of Origin: USA.
Type: Tactical transport helicopter.
Power Plant: Two 1,543 shp General Electric T700-GE-700 turboshafts.
Performance: Max. speed, 224 mph (360 km/h) at sea level; cruise, 166 mph (267 km/h); vertical climb rate, 450 ft/min (2,28 m/sec); hovering ceiling (in ground effect), 10,000 ft (3 048 m), (out of ground effect), 5,800 ft (1 758 m); endurance, 2·3 3·0 hrs.
Weights: Design gross, 16,500 lb (7 485 kg); max. take-off, 22,000 lb (9 979 kg).
Dimensions: Rotor diam, 53 ft 8 in (16,23 m); fuselage length, 50 ft 0¾ in (15,26 m).
Notes: The Black Hawk was winner of the US Army's UTTAS (Utility Tactical Transport Aircraft System) contest, and contracts had been announced by beginning of 1982 for 342 examples. The first of three YUH-60As was flown on 17 October 1974, and a company-funded fourth prototype flew on 23 May 1975. The Black Hawk is primarily a combat assault squad carrier, accommodating 11 fully-equipped troops, but it is capable of carrying an 8,000-lb (3 629-kg) slung load. Two variants under development at the beginning of 1981 were the EC-60A ECM model and EC-60B for target acquisition. The first production deliveries to the US Army were made in June 1979, with 225 delivered by beginning of 1982 against requirement for 1,107.

SIKORSKY S-70L (SH-60B) SEA HAWK

Country of Origin: USA.

Type: Shipboard multi-role helicopter.

Power Plant: Two 1,690 shp General Electric T700-GE-401 turboshafts.

Performance: (At 20,244 lb/9 183 kg) Max. speed, 167 mph (269 km/h) at sea level; max. cruising speed, 155 mph (249 km/h) at 5,000 ft (1 525 m); max. vertical climb, 1,192 ft/min (6,05 m/sec); time on station (at radius of 57 mls/92 km), 3 hrs 52 min.

Weights: Empty equipped, 13,678 lb (6 204 kg); max. take-off, 21,844 lb (9 908 kg).

Dimensions: Rotor diam, 53 ft 8 in (16,36 m); fuselage length, 50 ft 0¾ in (15,26 m).

Notes: Winner of the US Navy's LAMPS (Light Airborne Multi-Purpose System) Mk III helicopter contest, the SH-60B is intended to fulfil both anti-submarine warfare (ASW) and anti-ship surveillance and targeting (ASST) missions, and the first of five prototypes was flown on 12 December 1979, and the last on 14 July 1980. Evolved from the UH-60A (see page 247), the SH-60B is intended to serve aboard DD-963 destroyers, DDG-47 Aegis cruisers and FFG-7 guided-missile frigates as an integral extension of the sensor and weapon system of the launching vessel. The US Navy has a requirement for 204 LAMPS III category helicopters, with delivery of first seven in 1983, and 195 simplified SH-60Cs with deliveries from 1987.

SIKORSKY S-76

Country of Origin: USA.

Type: Fourteen-seat commercial transport helicopter.

Power Plant: Two 700 shp Allison 250-C30 turboshafts.

Performance: Max. speed, 179 mph (288 km/h); max. cruise, 167 mph (268 km/h); range cruise, 145 mph (233 km/h); hovering ceiling (in ground effect), 5,100 ft (1 524 m), (out of ground effect), 1,400 ft (427 m); range (full payload and 30 min reserve), 460 mls (740 km).

Weights: Empty, 4,942 lb (2 241 kg); max. take-off, 10,300 lb (4 672 kg).

Dimensions: Rotor diam, 44 ft 0 in (13,41 m); fuselage length, 44 ft 1 in (13,44 m).

Notes: The first of four prototypes of the S-76 flew on 13 March 1977, and customer deliveries commenced 1979, with 190 being delivered by the beginning of 1982, when a production rate of five per month was being maintained. The S-76 is unique among Sikorsky commercial helicopters in that conceptually it owes nothing to an existing military model, although it has been designed to conform with appropriate military specifications and military customers were included among contracts for some 400 helicopters of this type that had been ordered by the beginning of 1982. The S-76 may be fitted with extended-range tanks, cargo hook and rescue hoist. The main rotor is a scaled-down version of that used by the UH-60. Military customers for the S-76 include Jordan which has procured 16.

WESTLAND SEA KING

Country of Origin: United Kingdom (US licence).
Type: Anti-submarine warfare and search-and-rescue helicopter.
Power Plant: Two 1,660 shp Rolls-Royce Gnome H.1400-1 turboshafts.
Performance: Max. speed, 143 mph (230 km/h); max. continuous cruise at sea level, 131 mph (211 km/h); hovering ceiling (in ground effect), 5,000 ft (1 525 m), (out of ground effect), 3,200 ft (975 m); range (standard fuel), 764 mls (1 230 km), (auxiliary fuel), 937 mls (1 507 km).
Weights: Empty equipped (ASW), 13,672 lb (6 201 kg), (SAR), 12,376 lb (5 613 kg); max. take-off, 21,000 lb (9 525 kg).
Dimensions: Rotor diam, 62 ft 0 in (18,90 m); fuselage length, 55 ft 9¾ in (17,01 m).
Notes: The Sea King Mk 2 is an uprated version of the basic ASW and SAR derivative of the licence-built S-61D (see page 245), the first Mk 2 being flown on 30 June 1974, and being one of 10 Sea King Mk 50s ordered by the Australian Navy. Twenty-one delivered to the Royal Navy as Sea King HAS Mk 2s, and 15 examples of a SAR version to the RAF as Sea King HAR Mk 3s. Current production version is the Sea King HAS Mk 5 (illustrated), delivery of 17 to Royal Navy having commenced October 1980. All HAS Mk 2s will be brought up to Mk 5 standards. A total of 247 Westland-built derivatives of the S-61D had been ordered by the beginning of 1982.

WESTLAND WG 13 LYNX

Country of Origin: United Kingdom.
Type: Multi-purpose, ASW and transport helicopter.
Power Plant: Two 900 shp Rolls-Royce BS.360 07-26 Gem 100 turboshafts.
Performance: Max. speed, 207 mph (333 km/h); max. continuous sea level cruise, 170 mph (273 km/h); max. inclined climb, 1,174 ft/min (11,05 m/sec); hovering ceiling (out of ground effect), 12,000 ft (3 660 m); max. range (internal fuel), 391 mls (629 km); max. ferry range (auxiliary fuel), 787 mls (1 266 km).
Weights: (HAS Mk 2) Operational empty, 6,767–6,999 lb (3 069–3 179 kg); max. take-off, 9,500 lb (4 309 kg).
Dimensions: Rotor diam, 42 ft 0 in (12,80 m); fuselage length, 39 ft 1¼ in (11,92 m).
Notes: The first of 13 development Lynxes was flown on 21 March 1971, with the first production example (an HAS Mk 2) flying on 10 February 1976. By the beginning of 1982 production rate was nine per month and a total of 310 was on order, including 40 for the French Navy, 80 for the Royal Navy, 114 for the British Army, 10 for the Argentine Navy, eight for the Danish Navy, 12 for the German Navy, nine for the Brazilian Navy, six for Norway, 24 for the Netherlands Navy, three for the Nigerian Navy and three of a general-purpose version for Qatar. The Lynx AH MK 1 is the British Army's general utility version and the Lynx HAS Mk 2 is the ASW version for the Royal Navy. Eighteen of the Dutch and 14 of the French Lynx have uprated engines.

WESTLAND WG 30

Country of Origin: United Kingdom.
Type: Transport and utility helicopter.
Power Plant: Two 1,060 shp Rolls-Royce Gem 41-1 turboshafts.
Performance: Max. speed (at 10,500 lb/4 763 kg), 163 mph (263 km/h) at 3,000 ft (915 m); hovering ceiling (in ground effect), 7,200 ft (2 195 m), (out of ground effect), 5,000 ft (1 525 m); range (seven passengers), 426 mls (686 km).
Weights: Operational empty (typical), 6,880 lb (3 120 kg); max. take-off, 11,750 lb (5 330 kg).
Dimensions: Rotor diam, 43 ft 8 in (13,31 m); fuselage length, 47 ft 0 in (14,33 m).
Notes: The WG 30, flown for the first time on 10 April 1979, is a private venture development of the Lynx (see page 251) featuring an entirely new fuselage offering a substantial increase in capacity. Aimed primarily at the multi-role military helicopter field, the WG 30 has a crew of two and in the transport role can carry 17–22 passengers. Commitment to the WG 30 at the time of closing for press covers the flight development of two prototypes, and an initial production batch of 20 for delivery from mid-1982. British Airways has ordered two and a contract for six has been placed by the US-based Airspur Airline which also has options on a further 15. The WG 30 utilises more than 85% of the proven systems of the WG 13 Lynx, and the Mk 2 version will have the uprated Gem 60 turboshaft.

ACKNOWLEDGEMENTS

The author wishes to record his thanks to the many aircraft manufacturers that have supplied information and photographs for inclusion in this volume. The three-view silhouette drawings published in this volume are copyright Pilot Press Limited and may not be reproduced without prior permission.

INDEX OF AIRCRAFT TYPES